SO-BMW-732

Gift of the

SAMUEL P. HUNT FOUNDATION

Native American Mathematics

Native American Mathematics

Edited by Michael P. Closs

University of Texas Press, Austin

LIBRARY
COLBY-SAWYER COLLEGE
NEW LONDON, N.H. 03257

E
59
.M34
N37
1986

International Standard Book Number 0-292-75537-1
Library of Congress Catalog Card Number 86-50592

Copyright © 1986 by the University of Texas Press
All rights reserved
Printed in the United States of America

Third Printing, 1990

Requests for permission to reproduce material from this work should be sent to Permissions, University of Texas Press, Box 7819, Austin, Texas 78713-7819.

For reasons of economy and speed this volume has been printed from camera-ready copy furnished by the editor, who assumes full responsibility for its contents.

1495958

109776

Contents

Native American Mathematics

Preface

Modern mathematics is international in character. Its concepts are transmitted, studied and developed in numerous national languages in all parts of the world. The symbolic description of these concepts is presented in a universal mathematical notation independent of language. For example, as a part of this notation, numbers are expressed in a decimal system using Hindu-Arabic numerals.

The international nature of modern mathematics is a relatively recent phenomenon and represents a continuation of mathematical developments which occurred in Europe during the centuries from 1600 to 1900. The flowering of European mathematics was first nourished and stimulated through contacts with the Arabic world which had experienced an intellectual awakening during the great expansion of Islam. Arab savants had accumulated a repository of mathematical knowledge which drew on sources to be found in India, Persia and the Mediterranean world. These sources were themselves fed by the earlier mathematics of ancient Greece, Egypt and Babylonia. Thus, modern mathematics results from the cumulative effort of diverse peoples over thousands of years.

Historians of mathematics have concentrated on the great main stream leading to modern mathematics and have paid only scant attention, if any at all, to mathematics in cultures not directly contributing to it. There are exceptions to this tendency and some studies of Chinese, Japanese and African mathematics have appeared. In addition, some work has been done on the primitive origins of counting, arithmetic and geometry. The present volume is also exceptional in that it focuses on the mathematical development indigenous to the

New World. This is an area about which there is a dearth of information in the mathematical literature. It is my hope that this work will help to remedy this state of affairs and will lay a foundation for future studies in this area.

In my opinion, native American mathematics can best be described as a composite of separate developments in many individual cultures. The contributions to the volume are concerned with several aspects of this development among various native American groups. The papers, considered as a whole, give a good representation of the variety of mathematical experience found in the New World. The papers also give some idea as to the form which the history of mathematics must take if it is to incorporate material outside of its traditional boundaries. It is a form in which an almost total reliance on the historical approach is supplemented or replaced by drawing on the resources and methodologies of other disciplines such as anthropology, archaeology and linguistics.

ACKNOWLEDGEMENTS

I wish to thank the Canadian Society for the History and Philosophy of Mathematics for its early support of a project which led to the present volume. I am grateful to the Social Sciences and Humanities Research Council of Canada and the Rector's Fund of the University of Ottawa for providing me with research grants to conduct studies in this area over the past several years. I am indebted to Michelle Lukaszczyk, France Jean, and Madelaine Latour of the Department of Mathematics of the University of Ottawa for typing the copy for this work. Finally, I express my appreciation to the contributors to the volume whose patience and assistance have helped to make it a reality.

Michael P. Closs

1. Native American Number Systems

Michael P. Closs

NUMBER SYSTEMS

Many number systems of North and south America are decimal systems, or 10-systems, meaning that the formation of their number words is based on groupings of 10. This is true, for example, in the regions of North America occupied by the Algonquian, Siouan, Athapascan, Iroquoian and Salish linguistic stocks and in that part of South America dominated by the Quechua. However, in the Inuit area, most of Mexico and Central America, parts of California and the regions occupied by the Caddoan stock, the number systems are based on primary groupings of 20 and are called 20-systems. If there are secondary groupings of 5, or 10, we may further refine the nomenclature and speak of 5-20 systems, or 10-20 systems. In addition to these systems, there are some which use another base altogether or have no base at all.

It is worth mentioning that a diversity of number systems also existed outside of the Americas. Although most Old World peoples employed 10-systems, there were exceptions. The Celtic of northwestern Europe, the Ainu of northeastern Asia, the Yoruba, Igbo and Banda of Africa, and the native Australians of Victoria all used 20-systems. Moreover, many other native tribes of Australia, as well as the Bushmen of Africa, used 2-systems. Yet, this diversity is still not comparable to that found in the New World.

The numeral words up to ten are generally opaque in the Indo-European languages save for their numerical significance. The stems for these numerals are very uniform, even though they appear in languages which are mutually

unintelligible and so different that their common origin
would not be known but for study. Neither of these patterns
necessarily hold in other linguistic families. Indeed, many
Native American languages have numeral words below ten, which
illustrate digital origins or origins by arithmetical processes. And, while some families, such as the Mayan, exhibit
great uniformity in their numeral stems, others do not.
Perhaps the most extreme example of the latter is found among
the four languages of the Yukian family. With one exception
the numerals up to 3 in these dialects are related. From 4
on they differ completely and are all composite. In many
cases the meaning of the compositions is clear and it can be
seen that the significance of the numerals, the actions or
objects referred to, are almost invariably different. Moreover, even the methods of forming the numerals differ since
one of the four is an 8-system, two others are 5-10-systems
and the fourth is a 5-20-system (Dixon and Kroeber 1907,
p.670)

This paper begins with a general discussion of the origin
of native American number words and the problem of the upper
limit of counting in a number system. In the remainder of
the paper, the emphasis is placed on an examination of the
number systems from several specific cultural groups. This
will permit one to examine not only individual number words
but, more importantly, the systemic nature of numeral formation. The groups considered are selected to complement those
discussed elsewhere in the volume and are chosen to exemplify
the variety in the types of number systems found in the
Americas. They are presented in a more or less increasing
order of complexity.

ORIGIN OF NUMBER WORDS

The Inuit artists of Cape Dorset became involved with

printmaking in the late fifties. The documentation associated with the earliest prints was very casual and to remedy
this Dorothy Eber (1972) solicited further information about
these prints in the late sixties. The very first Dorset
print is by Pootagook, an Inuit leader, and is entitled
'Joyful I see Ten Caribou'. The print, depicts a grinning
Inuit with two upraised hands showing all ten digits. When
this was shown to Oshaweetok, who had participated in the
early print-making experiments, it evoked the following
response:

"It shows the indication for ten caribou ... this way of
counting came from the first Eskimo people. How did we count
to 100? We went by hands and then by feet. Two hands are 10
and one foot is 15. The other foot makes 20. When you have
20, that's one person. One person plus five fingers is 25
and so on. Five people make 100 and 100 means a bundle.
Often the foxes and sealskins were bundled into 100."

Oshaweetok's description clearly reveals a digital origin
for the Inuit method of counting. This is also demonstrated
by the following number terms in different Inuit dialects,
selected from Barnum (1901, pp.219-220), Eells (1913) and
Trumbull (1874).

Greenland	7	arfinek-mardluk,	'on the other hand, 2'
	9	mikkelerak,	'fourth finger'
	13	arkanenpingasut,	'on the first foot, 3'
Hudson's Bay	8	kittukleemot,	'middle finger'
	10	eerkitkoka,	'little finger'
Southwestern Alaska	5	tatlemen,	related to a term meaning

	10	koln,	'right hand' 'the upper half' (of the body)
	20	yuenok,	'man completed'
Unalit	11	atkahakhtok,	'it goes down' (from the hands to the feet)
	16	gukhtok,	'it goes over (to the other foot)

Trumbull and Eells give many other number words from native American languages whose origins are transparent. Among those exhibiting a digital origin are the following.

Massachusetts	1	pasuk,	from piasuk, 'very small'
Dakota	2	nonpa,	'to bend down'
Abnaki	3	nass,	expresses the idea of 'in the middle'
Massachusetts	3	nishwe,	from nashaue, 'halfway'
Pawnee	4	skitiks,	'fingers (of) hand'
Karankawan	5	natsa behema,	'one finger', i.e., the thumb
Ojibway	5	nanan,	'gone, spent', i.e., all the fingers
Hidatsa	5	kichu,	'completely turned down'
Klamath	5	tunep,	'away-hand'
Choctaw	5	tahlapi,	'the first (hand)

			finished'
Tano	6	manli,	'hand-piece', i.e., hand + piece of next
Klamath	6	nadshk-shapta,	'one I have bent down'
Omaha	7	penompa,	'finger-two'
Omaha	8	pethatbathi,	'finger-three'
Klamath	8	ndan-ksahpta,	'three I have bent over'
Cheyenne	9	sohhtu,	from na-so-toyos, 'my fourth finger'
(Southern) Wintun	10	pampa-sempta,	'two hands'
Maidu (Konkau)	10	ma-tsoko,	'hand doubled'
Navaho	20	natin,	from tine, 'man'
Shasta	20	tsec,	'man'
Tlingit	20	tlekha,	'one man'

Trumbull gives an unusual example of a word for 4 which has a digital origin but does not refer to man. He says that the Abipones of Paraguay express 4 as geyenknute, 'the ostrich's toes'.

The above data can be expanded considerably. The Atakapa of Texas (Gatschet and Swanton 1932) have:

9	woc icol han,	'without little finger'
10	woc pe,	'finish of the hands (or fingers)'.

The Yurok (Dixon and Kroeber 1907, p.684) have words for 7, 8 and 9 which are the names of the three middle fingers of the hand:

7	tserucek,	'pointer',i.e., the

		index finger
8	knewetek,	'long one', i.e.,
		the middle finger
9	qrerermeq,	'little finger'.

The Takelma of southwestern Oregon (Sapir 1903, pp.264-265) have:

6	haimis,	'in the hand-one
		(finger)'
7	haigam,	'in the hand-two'
8	haixin,	'in the hand-three'
9	haigo,	'in the hand-four'
10	ixdil,	'hands (both)'
20	yapamis,	'one man'
100	teimis,	'one male person'.

The Piro, an Arawakan group of eastern Peru (Matteson 1965, p.107) have:

5	pamyo,	'one hand'
6	patsruxire,	'another big
		finger'
7	payokhipre,	'another pointer
		rod'
8	yokhipre,	'pointer rod'
9	muturuxi,	'small finger'
10	pamole,	'one member-of-the-
		tribe'.

One of the most complete digitally based sequences from 1 to 10 is that of the Zuñi of New Mexico (Cushing 1892, pp.292-296). The Zuñi number words and their derivations, according to Cushing, are presented in Table 1.1. He says

the derivation of 5 as "the cut off" conveys either the meaning of the thumb since it is separated, or cut off, from the
other fingers or of 5 as the end, or cut off point, of Zuñi
counting at some earlier time.

Table 1.1.   The Zuñi number words from 1 to 10.

1	topinte	'taken to start with'
2	kwilli	'that (finger) put down with its like'
3	hai	'equally dividing one'
4	awite	'all (of the fingers)all but done with'
5	opte	'the cut off'
6	topalikya	'another brought to add with'
7	kwillilikya	'two brought to add with'
8	hailikya	'three brought to add with'
9	tenalikya	'all but all are brought to add with'
10	astemthla	'all of the fingers'

The Cupeño of southern California (Curtis 1926, p.177)
have a very explicit digital formation for the numbers from 5
to 10.   These are listed in Table 1.2.   The number words from
6 to 9 are distinguished by their length and in this regard
are probably only surpassed by number words generated in a
base 2 system.

Table 1.2.   The Cupeño numerals 1 to 10.

1	suplawut
2	wi
3	pa

4	wichu	
5	nu-ma-qananah	'my hand one-side'
6	nu-ma-qananah-suplawut-nu-ma-yahwanut	
		'my hand one-side one my hand other-side'
7	nu-ma-qananah-wi-nu-ma-yahwanut	
		'my hand one-side two my hand other-side'
8	nu-ma-qananah-pa-nu-ma-yahwanut	
		'my hand one-side three my hand other-side'
9	nu-ma-qananah-wichu-nu-ma-yahwanut	
		'my hand one-side four my hand other-side'
10	nu-ma-tolwunut	'my hand finished'

With respect to non-digital origins of number words, Eells says that the word for 1 has a connection with the first person pronoun in some languages. He observes that the word for 2 often comes from roots denoting separation or pairs. In this regard, Trumbull gives:

Micmac	2	tabu,	'equal'
Omaha	2	nomba,	'hands'
Apache	2	naki,	from ki-e, 'feet'.

The word for 3 sometimes has the meaning of 'more' or 'many'. An example of this appears in Micmac tchicht, 'three', which is cognate with Delaware tchitch, 'still more'.

Among the Yana of northern California (Sapir and Swadesh 1960) the word for 4 was daumi, apparently derived from dau, 'to count'.

A common non-digital method of forming number words is by

arithmetic principles. Unlike the situation in English, several native American languages use it to construct numbers less than 10. The additive principle is very often used to express 6, 7, 8, and 9 in the form 5 + 1, 5 + 2, 5 + 3, and 5 + 4. A supposedly rare phenomenon, namely the use of subtraction in forming number systems, was found in 40 percent of the more than 300 languages examined by Eells. As might be expected, "one subtracted" was the most frequent, occurring in 30 per cent of the languages, "two subtracted" in about 5 per cent, and "three subtracted" and "ten subtracted" in about 2 per cent each. As an example of "two subtracted", Eells cites the following example:

    Crow   8   nupa-pik, from upa, '2',and pirake, '10'.
Seidenberg (1960, pp.241-242) notes that the Yurucare of South America have 9 = 10 - 1, 8 = 10 - 2, and 7 = 10 - 3.

The multiplicative principle, like the additive is invariably used for the formation of higher numerals. In its simplest form, the duplicative, it is also frequently used for the formation of smaller numerals. Eells notes fifty cases of its use for the formation of 8, thirty-five for 4, twenty-five for 6, ten for 10, and two for 12. As examples, he includes:

Kutchin	6	neckh-kiethei,	from nackhai, '2', and kiethei, '3'
Kutchin	8	nakhai etanna,	from nackhai, '2', and etama, '3'
Kansas	8	kiya-tuba,	from kiya, 'again', and tuba, '4'
Gabrieleño	10	wehes-mahar,	from wehe, '2', and mahar, '5'
Cehiga	12	cape-nanba,	from cape, '6', and nanba, '2'.

The divisive principle is rarely used in the formation of numbers, the only examples found by Eells being:

Unalit	10	kolin,	'upper half of the body'
Pawnee	5	sihuks,	from ishu, 'hand',and huks, 'half', i.e., half of two hands.

To form the largest units in a given count notions of the superlative or the indefinite are sometimes employed. This is done, for example, in English where "thousand" is derived from Gothic pus-hundi, 'strong hundred', and "million" is derived from Italian milli-one, 'great thousand'. Examples provided by Eells are:

Delaware	1,000	ngutti kittapachki,	'great hundred'
Choctaw	1,000	tahlepa siponki,	'old hundred'
Kwakiutl	1,000,000	tlinhi,	'number which cannot be counted'.

To this we can add the Biloxi of Louisiana (Dorsey and Swanton 1912) who have 1,000 given by tsipitcya, 'old man hundred', and the Wiyot of northern California (Teeter 1964, p.93) who have 1,000 given by kucerawagatoril piswak, 'the counting runs out entirely once'.

The Fox (Jones and Michelson 1903, p.861) had two terms for 1,000, medaswakw, 'ten hundreds', and negutimakakw, 'one box'. The latter comes from negut, 'one', and makakw, 'box'. It is the more usual of the two and the more recent. Apparently in some of their earlier sales of lands to the government, the Fox received payment partly in cash. The money was brought in boxes, each box containing a thousand dollars. From this circumstance the term for "one box" entered the

language as an expression for 1,000 and managed to displace
the earlier more logical word.

Among the Osage (La Flesche 1932) we find 1,000 given by
zho-ku-ge, 'wooden box'. The derivation is similar to the
Fox and results from government payments to the Osage, in
silver dollars, which were packed in little wooden boxes,
each of which held one thousand dollars.

LIMITS OF COUNTING

The upper limits of counting vary considerably in different languages. According to Henry R. Schoolcraft (1851, II,
pp.204-221; V, pp.585-589), the Dakota, Cherokee, Ojibway,
Winnebago, Wyandot, and Micmac could all count into the millions, the Choctaw and Apache to the hundred thousands, and
many other tribes to 1000 or more. We also know that the
Aztec, Inca and Maya all counted into the millions. However,
this is not to say that there were in fact upper limits in
their counting.

For example, Bishop Baraga, (1878, pp.306-309) gives a
list of Ojibway number words, the largest being 1,000,000,
and then adds the term etc. implying that the number system
can be extended if desired. W. Warren (Schoolcraft 1851, II,
pp.211-213) makes this much more explicit and lists the following terms greater than or equal to 1,000,000:

1,000,000	me-das-wac da-sing me-das-wac
10,000,000	me-datch-ing me-das-wac
	me-das-wac
100,000,000	ningod-wac me-das-wac me-das-wac
1,000,000,000	me-das-wac me-das-wac as he
	me-das wac

He goes on to say that 1,000,000 is also called ke-che
me-das-wac, 'great thousand', which abbreviated the counting

a great deal. He adds that there "is no more limit (in thus counting) in the Ojibwa[y] then there is in the English language".

Diego de Landa (1941, p.98), the third bishop of Yucatan, expressed the same sentiment concerning the Maya number system. He put it nicely when he wrote: "They have other very long counts and they extend them in infinitum, counting the number 8,000 twenty times, which makes 160,000; then again this 160,000 by twenty, and so on multiplying by 20, until they reach a number which cannot be counted."

Some writers have expressed scepticism concerning the native ability to express such large numbers. For example, P. Prescott (Schoolcraft 1851, II, p.208), after listing Dakota number words as high as 1,000,000,000, comments:

"The Indians themselves have no kind of an idea what these amounts are; the only way they could form any kind of an idea would be to let them see the amount counted out. One thousand is more than or a higher number than some of them can count. We hear some of them talk about thousands, and sometimes a million, but still they can give no correct idea how much of a bulk it would make."

Prescott seems to assume that it is possible to conceive of large numbers in some direct sense. Clearly, one can visualize 1,000,000 as a symbol or "one million" as a word, but to suggest that one can comprehend the quantity is false. To give an idea of the "bulk" of a large number requires either familiarity with a collection of objects having the desired cardinality or an analytic technique in which a relatively small number of objects are placed in a group and then the "bulk" of this group is replicated by arithmetic principles until one attains a size sufficient to encompass the desired number of objects. The one notion reflects on the

utilitarian limits for number in a given culture whereas the
other reflects on the development of a culture's mathematical concepts.

Now, in general, number systems incorporate logical patterns determined by basic groupings and arithmetic principles
which enable the user to precisely quantify number in an
indirect manner. This ability is exactly what is required by
the so-called "analytic technique" referred to above. Thus,
if principles of grouping, addition, and multiplication are
implicit in the structure of a number system, Prescott's comments are rendered sterile. Number systems which can be
extended indefinitely, if required, do incorporate the above
principles and have attained a level of conceptual development which makes them equivalent to the set of positive
integers with the operations of addition and multiplication.

This is the situation with the Dakota, Ojibway, Maya, and
other groups mentioned above. Their number systems are
mature; the largest numbers actually used by these groups is
not really significant.

It is worth noting that even among groups in which the
number sequence does not extend beyond a thousand, there are
many which utilize the same arithmetic and grouping principles which occur in the mature number systems. These may
also be regarded as structurally complete. For example, in
many areas the traditional Inuit number system only reaches
into the hundreds; yet, structurally the systems are rich
enough that they could be extended were this desired.
Frances Barnum (1901, pp.219-220) does provide number words
for 1000 among the Inuit of south-western Alaska. He gives
taelemen epeat koloqkonut, 'five sets of twenty (taken) ten
times', or briefly 100 × 10, and also as an alternative
tesitsaq, a corruption of the Russian word for 1,000.

However, other reports are like that of Edward W. Nelson (1899, p.238) who observed that in the Bering Strait region most boys of 10 or 12 years of age could count objects very readily up to 100 and over and some men could reach 400. It would appear that in most cases a range of numerals less than 1,000 was adequate for the needs of the Inuit community. Perhaps, the relatively small numbers in actual use can also be partly traced to a negative attitude towards large numbers as revealed in the following folk tale of the Copper Eskimo.

Two hunters return, one with a wolf, the other with a caribou. They begin arguing as to which hide has the most hairs, and in order to settle the argument, decide to have a contest, each pulling the hairs out one at a time. They count and count and become so engrossed in what they are doing that days pass and they die of hunger. "That is what happens", the Eskimo storyteller adds, "when one starts to do useless and idle things that can never lead to anything." (Seidenberg 1962, p.33).

There is also a suspicion of larger numbers among the Crow. According to Eells (1913, p.298), they do not count above a thousand, as they say honest people have no use for higher numerals!

SIRIONA AND YANOAMA

The Siriona of Bolivia (Holmberg 1950, pp.47-48) and the Yanoama of the Amazonia in Brazil (Becher 1960, HRAF p.259) have the most rudimentary of number systems, each containing only three numbers. The Siriona count as in Table 1.3. Everything above 3 becomes etubenia, 'much', or eata, 'many'. The word for 3, yeremono, is clearly derived from the word for 2, yeremo.

Table 1.3.   The Siriona number sequence.

1   komi
2   yeremo
3   yeremono

The Yanoama employ the number words shown in Table 1.4.
Beyond 3 they say pruka, 'much', or pruka pruka, 'very much'.
These words are clearly related to the word for 3, prukatabo,
which in turn seems related to the word for 2, porokabo.

Table 1.4.   The Yanoama number sequence.

1   mahon
2   porokabo
3   prukatabo

The lack of number words greater than 3 does not indicate
that number sense breaks down above 3.   There still remains a
sense of whether a quantity has become larger or smaller,
though there is no way to express it numerically.   If something has been added to or subtracted from a set of objects
it may be detectable because of a change in the configuration
of the ensemble.   Indeed, Becher notes that "if 20 arrows are
standing together and one increases or reduces the bundle by
only one during the owner's absence, he will notice this
change at once upon his return."   In the same vein, Holmberg
asserts that: "A man who has a hundred ears of corn hanging
on a pole ... will note the lack of one ear immediately."
    It is also possible for a very precise notion of number
concept to exist independently of the use of number words.
This is illustrated by the Waica, a subgroup of the Yanoama,

who have words to designate 1 and 2. James Barker (1953,
HRAF p.57) writes that: "They can say two and one, or two and
two, but they generally do not do this. They show exact numbers higher than two by raising their fingers. I have seen
them crossing the dwelling to see if a person is holding up
three or four fingers in the semidarkness. I have asked for
as many as 12 objects and received the exact quantity by
showing them four fingers of my hand three consecutive
times."

BACAIRI

The Bacairi of Mato Grosso in Brazil (Steinen 1894, HRAF
pp.491-492, 506, 508) use the number sequence shown in Table
1.5. It can be seen that there are two number words for 3.
Steinen reports that the second, ahewao, is used no more
frequently than the form made up of 2 and 1. It is also
noteworthy that ahewao does not enter into any of the higher
number words, not even 6. This might lead one to think that
ahewao is a newer word but Steinen claims that linguistic
comparison with related tribes shows that it must be very
old, in fact, older than the word for 1. The word for 1 is
derived from the word for bow. It has been suggested that
since each man had only one bow but many arrows the bow came
to exemplify 'oneness'. The word for 2 and the word for
"many" derive from the same source. Thus, the two basic
words in the number vocabulary have a non-digital origin. It
is the word for 2 which has the greatest number of equivalents in related languages and Steinen sees this as evidence
that actual, conceptual counting began with 2.

Table 1.5. The Bacairi number sequence.

1   tokale

2	ahage	
3	ahage tokale ; ahewao	2 + 1 ; 3
4	ahage ahage	2 + 2
5	ahage ahage tokale	2 + 2 + 1
6	ahage ahage ahage	2 + 2 + 2

The Bacairi number sequence is formed by repetition of the words for 1 and 2. The largest grouping which is used is the 2-grouping. It is clear that the use of repetition in the extension of the sequence beyond 2 represents incipient addition. The formation of the numbers follows the pattern 1, 2, 2 + 1, 2 + 2, 2 + 2 + 1, 2 + 2 + 2. This is a typical example of an additive 2-system, that is, a system in which the number words are formed by using addition (implicitly) and grouping by 2's.

That the construction of the numbers involves 2-groupings and not simply a patterned ordering of two distinct words is made clear by an experiment of Steinen. He placed 3 maize kernels on the ground and asked for their number. Every informant first divided the pile into 2 and 1 before replying. The experiment was continued up to 6 kernels. It was observed that little heaps of 2 kernels were always made, they were always touched, and then finger computations were made before giving the result.

The Bacairi method of finger counting is also instructive. He starts with the little finger of the left hand and says tokale, grasps the adjacent finger and joins it with the little finger and says ahage, goes over to the middle finger and says, holding it separately beside the little finger and the ring finger, ahage tokale, goes over to the index finger, joins it with the middle finger and says ahage ahage, grasps the thumb and says ahage ahage tokale, places the little

finger of the right hand alongside it and says <u>ahage ahage</u>
<u>ahage</u>. Beyond 6 the Bacairi has come to the end of his
number words and now continues with the remaining fingers of
the right hand by touching each finger in turn and simply
adding <u>mera</u>, 'this one'. In like manner he touches the toes
of the left and right foot and each time says <u>mera</u>. If he is
still not finished, he grasps his hair and pulls it apart in
all directions.

It may be observed that after 6 there is an end to the use
of 2-groupings in finger counting. The finger counting then
climbs to 20 by a straight one-to-one correspondence without
the use of number words.

A number system, such as the Bacairi, which depends on
repetition for the formation of number words has an intrinsic
limitation. It cannot be extended very far before one loses
count of how many repetitions are involved.

THE BORORO

The Bororo, also of Mato Grosso, employed a 2-system which
developed into a vigesimal system based on the fingers and
toes. One version of their number sequence, reported by
General Rendon (Lounsbury 1978, p.761) is shown in Table 1.6.
The number words, with the exception of an alternative term
for 5 having a digital origin, clearly exhibit the structure
of a 2-system.

Fathers Colbacchini and Albisetti reported the same or
similar expressions, with the remark that ordinarily they
simply show the fingers of the hand, or of the two hands,
saying <u>inno</u> or <u>ainna</u>, 'thus', or <u>ainó-tujé</u>, 'only this many'.
They also noted that for 5 they show the left hand open,
saying <u>ikera aubodure</u>, 'my hand complete', that for 10 they
show both hands saying <u>ikera pudugidu</u>, 'my hand together',

Table 1.6. The Bororo number sequence (version 1).

1	mit	
2	pobo	
3	augere pobe ma awo metuya bokware	2 + 1
4	augere pobe augere pobe	2 + 2
5	augere pobe augere pobe awo metuya bokware	2 + 2 + 1
5	awo kera upodure, 'this my hand all together'	
6	augere pobe augere pobe augere pobe	2 + 2 + 2
7	augere pobe augere pobe augere pobe awo metuya bokware	2 + 2 + 2 + 1
8	augere pobe augere pobe augere pobe augere pobe	2 + 2 + 2 + 2
9	augere pobe augere pobe augere pobe augere pobe awo metuya bokware	2 + 2 + 2 + 2 + 1
10	augere pobe augere pobe augere pobe augere pobe augere pobe	2 + 2 + 2 + 2 + 2

and that going beyond 10 they employ the toes of one foot, and beyond 15 the toes of the other foot. When the objects counted are more than a number that they can express easily, they say makaguraga, 'many', or makaaguraga, 'very many' (Lounsbury, 1978, p.761).

In the summer of 1950, Floyd G. Lounsbury (1978, p.761) obtained another version of the Bororo number sequence, partially shown in Table 1.7. He observed that the number words were given as if they were names for numerals, without accompanying gestures of finger showing or pointing to toes. Lounsbury contrasts the long descriptive phrases in the Bororo system with more advanced systems such as the Mayan in which the words for the numerals up to 10 consist of brief unanalyzable roots. He attributes this difference in the development of a number vocabulary to a "long history of ellipses, of accumulation of phonetic changes with resulting loss of etymological transparency, and of repeated abbreviations accompanying increased frequency of use".

It is apparent from the two versions of the Bororo number sequence that the number systems as recorded were in transition from a 2-system to a 2-20-system. Indeed, in the first version there was already an alternative term for 5 having a digital origin. In the second version all the terms after 4 refer to the digits of the body or to natural groups of such digits although the pairing principle is still reflected in the term for 7.

It is interesting to note a report of Steinen (1894, HRAF p.652) that when counting kernels of maize the Bororo, like the Bacairi, first made little heaps of two and then examined their fingers before giving the total. Thus, the counting in the presumably older 2-system was also digitally based. This is apparent from both the gesture numbers mentioned by

Table 1.7. The Bororo number sequence (version 2).

1   ure mitótuǰe,   'only one'
2   ure póbe,   'a pair of them'
3   ure póbe ma ǰéw metúya bokwáre,   'a pair of and that
        one whose partner is lacking'
4   ure póbe púibỉǰi,   'pairs of them together'
5   ure ikéra aobodúre,   'as many of them as my hand
        complete'
6   bóture ikéra aobowúto,   (something about changing to
        the other hand)
7   ikéra metúya pogédu,   'my hand and another with a
        partner'
8   ikerakó boeyadadáw,   'my middle finger' (on the second
        hand)
9   ikerakó boeyadádaw mekíw,   'the one to the side of my
        middle finger'
10 ikerakó boeǰéke,   'my fingers all together in front'
13 ičare búture ivúre boeyadadawúto pugéǰe,   'now the one
        on my foot that is in the middle again'
15 ičare ivúre iyádo,   'now my foot is finished'
20 avúre ičare maka réma avúre,   'your feet, now it is
        as many as there are with your feet'
21   otúre turegodáǰe pugéǰe,   'starting them over again'
22   ure póbe turegodáǰe pugéǰe,   'two of them, starting
        over again'

Colbacchini and Albisetti and Steinen's finger consultation
based on pairing. Because of the digital basis underlying
the 2-system it would be natural to replace the long 2-system
terms for 5 and 10 by more convenient terms related to the
grouping of digits by hands. The extension of the system to
include the digits of the feet which leads ultimately to a
vigesimal system would proceed in a natural manner.

TOBA

The number system of the Toba of Paraguay (McGee 1900,
p.838), recorded in Table 1.8, reaches to 10. It has separate number words for only 1, 2, and 4, the remaining words
being composites of these. A peculiar feature of the system
is the presence of two distinct words for 2, both used in the
construction of the higher numerals. What guides the selection of a particular 2 in a given composite is puzzling. It
may be that the alliterative formation for the multiplicative
composites 6 = 2 × 3 and 8 = 2 × 4 and the contrary formation
for the additive composites 5 = 2 + 3 and 10 = 2 + (2 × 4) is
intentional. In this regard, it should be noted that the
number 1 is always used additively.

Table 1.8.   The Toba number sequence.

1	nathedac	
2	cacayni, nivoca	
3	cacaynilia	2 + lia
4	nalotapegat	equals (they) say
5	nivoca cacainilia	2 + 3
6	cacayni cacaynilia	2 × 3
7	nathedac cacayni cacaynilia	1 + (2 × 3)
8	nivoca nalotapegat	2 × 4

9	nivoca nalotapegat nathedac	$(2 \times 4) + 1$
10	cacayni nivoca nalotapegat	$2 + (2 \times 4)$

The Toba word for 3 is derived from the word for 2, as was the case with the Siriona. The word for 4 is derived from a word meaning "equals" and implicitly suggests the notion of a 2-grouping, though the concept does not appear explicitly in the formation of the other numbers. The numbers above 4 are expressed by using additive and multiplicative principles. This early use of arithmetic principles in the development of the number sequence contrasts with English usage where such principles only become evident in numbers above 10.

While the multiplicative principle is present in the Toba number system, it should be realized that it only occurs in its simplest guise, that of doubling. Nevertheless, it is significant that the arithmetic concepts of addition and multiplication are already incipient in such a limited number system.

COAHUILTECAN

The Coahuiltecan of Texas (Eells 1913, pp.268, 297; Thomas 1900, p.881), also referred to as the Rio Norte and San Antonio of Texas (Gallatin 1845, Table A), have a number sequence which, like the Toba, exhibits an extensive use of additive and multiplicative principles early in its development. Many of the Coahuiltecan number words are given in Table 1.9. The formation of 3 as an additive composite of 2 and 1 recalls the formation of 3 in the 2-systems considered earlier. However, this pattern is not continued and new words are introduced for 4 and 5. The number 6 is expressed in two ways, either by another new word or as a composite of 3 and 2. The sequence is extended to 10 and beyond by

Table 1.9. The Coahuiltecan number sequence

1	pil	
2	ajtê	
3	ajti c pil	2 + 1
4	puguantzan	
5	juyopamáuj	
6	ajti c pil ajtê ; chicuas	(2 + 1) × 2
7	puguantzan co ajti c pil	4 + (2 + 1)
8	puguantzan ajtê	4 × 2
9	puguantzan co juyopamauj	4 + 5
10	juyopamauj ajtê	5 × 2
11	juyopamauj ajtê co pil	(5 × 2) + 1
12	puguantzan ajti c pil	4 × (2 + 1)
13	puguantzan ajti c pil co pil	4 × (2 + 1) + 1
14	puguantzan ajti c pil co ajtê	4 × (2 + 1) + 2
15	juyopamauj ajti c pil	5 × (2 + 1)
16	juyopamauj ajti c pil co pil	5 × (2 + 1) + 1
17	juyopamauj ajti c pil co ajtê	5 × (2 + 1) + 2
18	chicuas ajti c pil	6 × (2 + 1)
19	chicuas ajti c pil co pil	6 × (2 + 1) + 1
20	taiguacô	
30	taiguacô co juyopamauj ajtê	20 + (5 × 2)
40	taiguacô ajtê	20 × 2
50	taiguacô ajtê co juyopamauj ajtê	(20 × 2) + (5 × 2)

arithmetic operations until 20 is reached and another new
term is introduced. Higher numerals are formed by building
on the base of 20 and using the same arithmetic principles as
occur earlier in the sequence. It can be seen that the multiplicative principle is more fully developed than in the
Toba system. It is not limited to simple doubling as is
made clear by the composites for 12, 15, and 18.

The word for 20 is not a composite and is used as a base
for constructing 30, 40, and 50. In order to rise from 12 to
20 there is a regular use of 3-groups which is very unusual
in number systems. Below 12 the development of the number
words is not consistent and one can find assorted groupings
of 2, 3, 4, 5 and 10. Were it not for the somewhat ambiguous
situation below 12, one could classify the Coahuiltecan
sequence as a 2-3-20-system.

YUKI

The Round Valley, or Yuki proper, dialect is one of four
Yukian languages, all of which are found in California (Dixon
and Kroeber 1907, pp.677, 684-685). The Yuki method of
counting is presented in Table 1.10 together with an analysis
of the numerals. It can be seen from the table that the Yuki
counted in groups of 8. Dixon and Kroeber erroneously classified the Yuki system as a 4-system in their work of 1907.
It was reclassified as an 8-system in a later publication of
Kroeber (1925, pp.176-177, 878-879). The analysis of the
numerals in Table 1.10 is consistent with the information in
the two preceding references but is somewhat different from
the analyses they contain so as to better illustrate the
nature of the number system.

This system of counting is inextricably associated with
the Yuki method of finger counting. Rather than counting the

Table 1.10. The Yuki (Round Valley) number sequence.

1	pa-wi	
2	op-i	
3	molm-i	
4	o-mahat, op-mahat	'two-forks'
5	hui-ko	'middle-in'
6	mikas-tcil-ki	'even-tcilki'
7	mikas-ko	'even-in'
8	paum-pat	'one-flat'
9	hutcam-pawi-pan	'beyond-one-hang'
10	hutcam-opi-sul	'beyond-two-body'
11	molmi-sul	'three-body'
12	omahat-sul	'four-body'
13	huiko-sul	'five-body'
14	mikastcilki-sul	'six-body'
15	mikasko-sul	'seven-body'
16	hui-co(t), '8'	'middle-none', '8'
17	pawi-hui-luk, '9'	'one-middle-project', '9'
18	opi-hui-luk, '10'	'two-middle-project', '10'
19	molmi-hui-poi, '11'	'three-middle-project', '11'
20	omahat-hui-poi, '12'	'four-middle-project', '12'
24	'8'	'8'
26	'10'	'10'
35	'19'	'19'
51	'19'	'19'
64	omahat-tc-am-op	'four-pile-at'

fingers themselves they counted the spaces between them, in each of which, when the manipulation was possible, two twigs were laid. Except for the words for 1 and 2, common to all the Yukian languages, and the word for 3, common to all but the Wappo, the Yuki number words are descriptive of this process of counting and have no relation to the number words used in the other related languages. From 9 to 15 the number words are formed by addition to a base of 8. In particular, from 10 to 15 the number words include the term sul, 'body', suggesting that 'body' represents a full count of the spaces between the fingers. Since the number words above 3 refer to the method of counting rather than to elements of a formal number sequence it is not surprising that sometimes there are several ways of denoting a number. For example, 8 is variously '1-flat', 'hand-stick-flat', or 'hand-2-only'.

An examination of the available expressions for numbers above 15 reveals that the following residue representations are possible:

$$16 = (8) + 8$$
$$17 = (8) + 9$$
$$18 = (8) + 10$$
$$19 = (8) + 11$$
$$20 = (8) + 12$$
$$24 = (2 \times 8) + 8 = (16) + 8$$
$$26 = (2 \times 8) + 10 = (16) + 10$$
$$35 = (2 \times 8) + 19 = (16) + 19$$
$$51 = (4 \times 8) + 19 = (2 \times 16) + 19$$

This indicates that 16, as well as 8, may be used as a base for constructing the higher numbers. Moreover, the term for 64 is literally '4-pile-at' and since 64 = 4 × 16 we may wonder if perhaps the counting is not proceeding by piles of 16 rather than groups of 8. Kroeber states that 64 is used as a higher unit in the Yuki count and thus credits the Yuki with a pure 8-system. However, on the basis of the evidence considered it seems more likely that the Yuki had evolved an 8-16-system.

The Yuki example illustrates how completely this system, many of whose terms do have reference to the fingers, departs from the more common modes of counting. It provides a wonderful demonstration that finger counting does not always lead to counting by fives and tens, a lesson which could also have been drawn from the finger counting of the Bacairi, considered in a previous section.

It is interesting to observe that the Yuki had precise concepts of number and counting which went far beyond their formal number sequence. These concepts provided the mechanism to extend the number sequence above 3, a process which was not yet formalized at the time the system was recorded. This can be seen by the existence of variant terms for the same number and the use of residue expressions for larger numbers.

MAIDU

The most common method of counting involves addition to the highest multiple of the base below the given number. For example, when we say 78 we understand 70 + 8, 70 being the highest multiple of 10 below the number 78. Menninger (1969, pp.76-80) refers to this as 'counting from the lower level'. An alternate method of counting is to count towards the next higher level. This procedure he refers to as 'overcounting'. Menninger regards overcounting as a rare phenomenon and discusses its occurrence in the Germanic north of Europe as well as among the Maya and the Ainu. Overcounting also occurs among the Maidu of California where it is even used in the

Table 1.11.   The Northwestern Maidu number sequence
             (Konkau dialect).

1	wikte	
2	pene	
3	sapu	
4	tsoye	
5	ma-tsani	'hand-tsani(?)'
6	sai-tsoko	'3-double'
7	matsan-pene	'5 + 2'
8	tsoye-tsoko	'4-double'
9	tsoye-ni-masoko	'4 with 10','4 towards 10'
10	ma-tsoko	'hand-double'
11	wikem-noko	'1-arrow'
12	pene-wikem-noko	'2-1-arrow'
13	sapwi-ni-hiwali	'3 with 15'
14	tsoye-ni-hiwali	'4 with 15'
15	hiwali	
16	wok-ni-maiduk-woko	'1 with man-1'
17	peni-maiduk-woko	'2 with man-1'
18	sapwi-ni-maiduk-woko	'3 with man-1'
19	tsoye-ni-maiduk-woko	'4 with man-1'
20	maiduk-woko	'man-1'
30	matsok-ni pene-ma	'10 with 2-man'
40	peni-ma	'2-man'
50	matsok-ni spawi-ma	'10 with 3-man'
60	sapwi-ma	'3-man'
70	matsok-ni tsoye-ma	'10 with 4-man'
80	tsoye-ma	'4-man'
90	matsok-ni matseni-ma	'10 with 5-man'
100	matseni-ma	'5-man'

formation of numerals below 20.

The number sequence of the Northwestern Maidu, Konkau
dialect (Dixon and Kroeber 1907, pp.679, 687-688) is listed
in Table 1.11. The frequent -ni-, 'with', in the Maidu list
is to be taken as signifying 'towards'. Thus the word for
13, sapwi-ni-hiwali, '3 with 15', means '3 towards 15'
(counted from the preceding level of 10). The use of overcounting first appears in the word for 9 which means '4
towards 10' and is regularly used from 13 up.

The additive principle is used explicitly only in the formation of 7 but appears implicitly wherever overcounting is
used. The multiplicative principle is used in the formation
of 6, 8, and 10 in the sense of doubling, and in the multiples of 20 in the sense of counting base units. The overall
structure is clearly that of a 5-20-system.

There is much confusion concerning the words for 11 and
12. Dixon and Kroeber interpret the word for 11, wikem-noko,
'1-arrow', as implying 1-arrow = 11. But then the word for
12, pene-wikem-noko, '2-1-arrow' or '2-11', would not be sus-
ceptible to a reasonable arithmetical explanation. Moreover,
among the Northwestern Maidu of Mooretown one finds 11,
wikte-ni wikem-noko, '1-with-1-arrow', and 12, wokem-noko,
'1-arrow', implying 1-arrow = 12. To add to the confusion,
the Mooretown dialect also has 20, penin nokom, '2-arrow',
implying 1-arrow = 10. The Northwestern Maidu near Chico
counted from 1 to 20 like the Konkau, with the exception of
11 and 12 which were expressed by wik-ni hiwali, '1 with 15',
and pe-ni hiwali, '2 with 15', respectively. In a later
record of the Northwestern Maidu, referred to as Valley Maidu
(Curtis 1924, p.234), the sequence is essentially the same as
the Konkau but again with the exception of 11 and 12 where
one has wuk-ni peneke, '1 with 12' and pene-ke, '12',

respectively.

LUISEÑO

Luiseño is one of the dialects belonging to the southern California branch of Shoshonean, a subfamily of Uto-Aztecan. The Luiseño (Kroeber and Gracea 1960, pp.118-121) have basic numerals only up to 5. In addition to cardinal forms, these numbers also take four other forms corresponding to plural, distributive, multiplicative and ordinal numbers. Cardinals above 5 are expressed by a variety of descriptive phrases exhibiting digital or arithmetical origins. There are often several ways of expressing such numbers. Some of these expressions and the five basic numerals are shown in Table 1.12.

Two patterns are apparent in the number formations above 5. In one pattern, numbers are formed using the structural principles of a pure 5-system. A beautiful example of this is the expression for 71 which is equivalent to

$$2 \times (5 \times 5) + (4 \times 5) + 1.$$

This method only seems to appear in numbers below 100. The second pattern uses the principles of a 5-20-system, with a group of 20 being represented by the phrase 'all my-hand my-foot finished'. In both types of numeral construction, multiplication and addition are freely used. However, the numbers used as multipliers are restricted to the five numbers of the formal number sequence.

It is interesting to note that while there are only five basic words used for counting, there are unit terms for 6,7, and 8 used in gambling. Thus the formal number sequence used for gambling is more extensive than that for counting.

The Luiseño example shows that precise number concepts reaching into the low hundreds and well utilized principles

LIBRARY
COLBY-SAWYER COLLEGE
NEW LONDON, N.H. 03257

109776

Table 1.12. Some Luiseño numerical expressions.

1	supul
2	wex
3	pahi
4	wasa
5	mahar
6	'again 1', 'another besides 1' 'five one upon' 'passing-over to-my-hand to-one finger'
8	(like 6 with 'three' substituted for 'one')
10	'my-hand finished both' 'all my hand finished' (like 6 with 'five' substituted for 'one')
11	'twice five one upon' 'besides other my-hand one finger'
15	'all my-hand finished and one my-foot'
16	'besides my-foot one digit (toe)' 'thrice five one upon'
20	'another finished my-foot the-side' 'four-times five'
21	'besides other my-foot one finger'
25	'all my-hand my-foot finished and another five'
30	'five-times five, five upon'
40	'twice my-hand my-foot finished' 'all my-hand my-foot finished again all my-hand my-foot finished'
71	'five-times five another five-times five, and four-times five, one upon'
80	'four-times all my-hand my-foot finished'
100	'five-times all my-hand my-foot finished'
200	'again five-times all my-hand my-foot finished'

of multiplication and addition are possible in a number system in which the formal part of the number sequence consists
of only five elements.

POMO

The Pomo of California have a deserved reputation as great
counters. Large counts were commonly performed by the Pomo
at the time of deaths and peace treaties. An example of such
a count appears in a Pomo tale which relates that the first
bear shaman gave 40,000 beads in pretended sympathy for the
victim whose death he had caused. Edwin Loeb (1926, p.230)
reports that his informant has observed counting in excess of
20,000. Kroeber (1925, pp.256-257; 879) believes that the
Pomo interest in counting developed from the wealth they
acquired by being the principal purveyors of the standard
disk currency to north-central California and the experience
thereby gained in counting long strings of clam-shell beads.

Although the Pomo were able to express numbers reaching
into the thousands, Dixon and Kroeber (1907, pp.676, 685-686)
only list their numerals up to 200. They also write that
"all the systems are entirely quinary-vigesimal, except the
Southeastern, which while decimal above ten is largely borrowed from the neighboring Wintun, and the Southern dialect,
which is decimal from forty up." This comment is not completely justified, as can be seen from Tables 1.13 and 1.14
illustrating Dixon and Kroeber's numeral lists for the Eastern and Southwestern dialects. It is clear that while the
Eastern Pomo used a 5-(10)-20-system, the Southwestern Pomo
used a 5-(10)-40-system.

The middle columns of the tables give analyses of the
numeral formations. These are connotative rather than etymological translations and, with the exception of Southwestern

Table 1.13.   The Pomo (Eastern) number sequence.

1	kali		
2	xotc		
3	xomka		
4	dol		
5	lema		
6	tsadi	1-di	(5) + 1
7	kula-xotc	kula-2	(5) + 2
8	koka-dol	2-ka-4	2 × 4
9	hadagal-com	10-less	10 - (1)
10	hadagal-tek	10-full	10
11	hadagal-na-kali	10 + 1	10 + 1
12	hadagal-na-xotc	10 + 2	10 + 2
13	hadagal-na-xomka	10 + 3	10 + 3
14	xomka-mar-com	3-mar-less	15 - (1)
15	xomka-mar-tek	3-mar-full	3 × 5
16	xomka-mar-na-kali	3-mar + 1	15 + 1
17	xomka-mar-na-xotc	3-mar + 2	15 + 2
18	xomka-mar-na-xomka	3-mar + 3	15 + 3
19	xai-di-lema-com	stick-di-5-less	20 - (1)
20	xai-di-lema-tek	stick-di-5-full	20
21	xai-di-lema-na-kali	stick-di-5 + 1	20 + 1
30	na-hadagal	na-10	(20) + 10
40	xotsa-xai	2-stick	2 × 20
50	hadagal-e-xomka-xai	10-e-3-stick	10 to 60
60	xomka-xai	3-stick	3 × 20
70	hadagal-ai-dola-xai	10-ai-4-stick	10 to 80
80	dol-a-xai	4-stick	4 × 20
90	hadagal-ai-lema-xai	10-ai-5-stick	10 to 100
100	lema-xai	5-stick	5 × 20
200	hadagal-a-xai	10-stick	10 × 20

Table 1.14.   The Pomo (Southwestern) number sequence.

1	ku		
2	ko		
3	sibo		
4	mitca		
5	tuco		
6	lan-tca	lan-1	(5) + 1
7	lan-ko	lan-2	(5) + 2
8	komtca	2-4	2 × 4
9	tcatco	1-tco	(10) - 1
10	tca-coto	1-10	10
11	nan-tca	+ 1	(10) + 1
12	na-ko	+ 2	(10) + 2
13	na-sibo	+ 3	(10) + 3
14	si-hma-con	3-hma-less	15 - (1)
15	si-hma-tek	3-hma-full	3 × 5
16	si-hma-nan-tca	3-hma + 1	15 + 1
17	si-hma-na-ko	3-hma + 2	15 + 2
18	si-hma-na-sibo	3-hma + 3	15 + 3
19	tca-hma-con	4-hma-less	20 - (1)
20	tca-hma	4-hma	4 × 5
21	tca-hma-nan-tca	4-hma + 1	20 + 1
30	lantca-hma	6-hma	6 × 5
40	ku-hai	1-stick	1 × 40
50	coto-hma	10-hma	10 × 5
60	si-hmak-tcidu	3-hma-tcidu	15 × (4)
70	si-hmak-tcidu-tcacoto	3-hma-tcidu-10	60 + 10
80	ko-hai	2-stick	2 × 40
90	ko-hai-tcacoto-ko	2-stick-10-ko	80 + 10
100	ko-hai-tcahma-ko	2-stick-20-ko	80 + 20

100, follow Dixon and Kroeber. In Southwestern 100, ko-hai-
tcahma-ko, I have taken tcahma to be the word for 20; this
leads to the analysis "2-stick-20-ko". The right hand columns of the tables give second order arithmetical analyses so
as to better indicate the structure of the number sequences.

In addition to the different bases employed in the two
dialects, it is interesting to note that one employs overcounting while the other does not. For example, 70 in the
Eastern dialect is '10 towards 80' whereas in the Southwestern dialect it is '60 + 10'.

The lack of detailed information concerning the larger,
numerical terms of the Pomo has been remedied by Loeb (1926,
pp.229-230). He provides the following sequence for large
counts among the Eastern Pomo:

80	dol-a-xai	4 sticks
100	lema-xai	5 sticks
200	hadagal-a-xai	10 sticks
300	xomka-mar-a-xai	15 sticks
400	kali-xai	1 (big) stick
500	kali-xai-wina-lema-xai	400 + 5 sticks
800	xote-guma-wal	2 (big sticks)
2400	tsadi	6 (big sticks)
3600	hadagal-com	9 (big sticks)
4000	hadagal	10 (big sticks)

The number words of the Eastern Pomo are closely connected
with their method of counting. The word for 20 is called
xai-di-lema-tek, 'full stick', and in counting small amounts
a stick is laid out for this primary unit. It is apparent
from the number sequence that when 20 such sticks were

accumulated they formed a larger unit of 400, also represented by a stick.

Loeb describes two methods of counting which were used when large quantities of beads were involved. He writes:

"According to the first, and older method, a small stick is laid out for every eighty beads. When five of these small sticks have been laid out, they are taken back, and a larger stick substituted for the Pomo large unit of four hundred. According to the second method a small stick is laid out for every hundred beads, four of these small sticks making the large unit. When four hundred has been reached the counting goes on in units of four hundreds until ten of the larger sticks have been used and four thousand beads have been counted. Now another group of ten sticks is prepared. They are all equal in size, a little larger than the former bundle of ten, and have some mark to distinguish them. Each stick represents four thousand. It must not be forgotten that while you are going on with each of these counts the previous bundle must be counted before you can 'put out a new stick'. Hence a great number of sticks are in use at one time. When each of these latter ten sticks have been counted, you reach the number forty thousand, xai-di-lema-xai. This is known as the 'big twenty'."

It can be seen that in the older method referred to above, an 80-count is used to reach 400, while in the other method a 100-count is used. Neither of these methods of grouping is consistent with the Eastern sequence and its 20-count or the Southwestern sequence and its 40-count. Dixon and Kroeber's list for the Southern Pomo shows that it is decimal after 40, with 100 expressed by '10-stick' and 200 by '2-stick'. Thus, in the Southern dialect one does find a 100-count represented by sticks. The fact that 1-stick represents counts of 20,

40, 80, and 100 in different sequences need not be regarded
as contradictory but only as evidence that variant usages
prevailed at different times and places in the Pomo area.
The use of 1-stick to represent smaller and larger groupings
in the same sequence is justifiable since the terms refer to
physically differentiable sticks. It is probable that 400,
as a higher unit, is common to all the Pomo sequences. The
word for 40,000, xai-di-lema-xai, appears to signify '(1 big)
stick times 5 sticks', that is, '400 x 100'.

The surprising flexibility in Pomo counting below 400 is
derived from the methods of stick counting. The various
numeral vocabularies result from the size of the groups
represented by 1-stick and so it is easy to see how several
distinct sequences were built up. One of the consequences of
this variation is that the Pomo number system does not have a
unique classification; it is 5-20-400, 5-40-400, (5)-80-400
or (10)-100-400.

The Pomo kept a record of the number of days which it took
to make a journey by tying knots in a string (Loeb 1926,
p.231). Each knot stood for a day's travel, the knots being
tied at night while the travellers were in camp. A record of
this kind usually contained only four or five knots and was
called kamalduyik, 'day count'.

Loeb (1926, p.231) also describes the Pomo method of
keeping records of wampum exchanged at a treaty. He writes:

"Supposing that one group gave the other a peace offering
of say twenty-four thousand wampum. The group which received
the wampum made a record of the number of beads received. To
do this they tied one knot in a string for every four hundred
wampum received. This was used as a check on the tying
together of the counting sticks. When sticks were used a
certain small stick indicated one hundred wampum, and a

larger stick four hundred wampum. The smaller stick measured
one and three-quarters inches [4.5 cm] in length and the
larger stick about three inches [7.5 cm]. The string record
and the stick record were put away together in a bag and kept
until the party returned the wampum to the original donors."

CONCLUSION

The number systems examined in the preceding sections
exhibit different levels of maturity, variety in their numerical bases, distinctive processes of numeral formation, and
also share a more or less contemporaneous existence. I think
that they provide clear evidence of the local generation of
number sequences. This can be seen in the occurrence of rare
features such as the peculiar significance of the number 3 in
Coahuiltecan counting, the unusual 8-16 number system of the
Yuki, and the repeated emphasis on the number 5 in Luiseño
numerical expressions. It can also be seen in the extension
of native number systems in a local context. For example,
Bororo counting evolved from a 2-system into a 2-20 system.
Even though the impetus for this development may have origi-
nated through contact with the dominant European cultures,
the underlying formative principles appear to have been a
proper part of Bororo culture from the earliest times. The
growth of a number system can also be seen in the various
methods of stick counting employed by the Pomo. It can be
argued that the idea of stick counting may have been transmitted to the different Pomo groups through a process of
diffusion. Yet, the application of stick counting to the
generation of large numerals proceeded in very independent
ways in the different Pomo groups. The fact that the stick
numerals were not standardized suggests that they may have
been a recent innovation in which local creativity surpassed

the transmission of linguistic terminology. A somewhat
similar phenomenon can be seen in Luiseño where one finds
fixed number words only up to 5. Beyond that one encounters
a creative and non-standardized descriptive terminology. In
the small, the local generation of number words can be seen
in the different terms which appear for 11 and 12 in the
various Maidu dialects and in some of the more unusual number
terms encountered earlier, such as the Abipones word for 4
given by geyenknute, 'the ostrich's toes'.

In the Americas, there was a widespread coexistence of
native number systems with different number bases and with
different methods of forming number words. While most native
languages employ additive and multiplicative principles in
numeral formation, it is only some which employ the subtrac-
tive and divisive principles. Again, while some languages,
like Toba and Coahuiltecan, exhibit a considerable reliance
on arithmetical principles in the formation of even small
numbers, most do not. In addition, it may be recalled that
although many languages, like the Southwestern Pomo, employ
'counting from a lower level', it is only some, like the
Maidu and the Eastern Pomo, which employ 'overcounting'.
Finally, while many languages utilize finger counting, they
do so in widely different ways as can be seen by comparing
the practices of the Bacairi and Bororo (2-groupings), the
Yuki (4-groupings), and the Maidu (5-and 20-groupings). Such
systemic differences describe an environment in which the
contrastive nature of local traditions is evident. They
reveal a cultural mosaic in which independent invention could
and apparently did flourish.

I am of the opinion that the early settlers of the New
World may well have brought number systems, possibly but not
necessarily rudimentary, from the Old World. However, as the

bands of immigrants spread and dispersed throughout the Americas, there must have been a continual fragmentation and reorganization of the bands which ultimately gave rise to the numerous linguistic and cultural groups found at the present time. In this process there is certainly room for varied levels of preservation of numerical concepts. The proposals of A. Seidenberg (1974; 1976), insofar as they relate to the Americas, give evidence of such a diffusion. Nevertheless, given the diversity found in the number systems of the New World, I would argue that not only was there diffusion, but above all, there was a loss and regeneration of counting skills which took place over and over again. This invention or re-creation of number systems would rarely have been without any foundation at all. Diffusion would still play a role. However, the evidence which has been presented indicates that independent invention is a primary ingredient in the development of many native number systems.

ACKNOWLEDGEMENTS

This work has been supported by a research grant from the Social Sciences and Humanities Research Council of Canada (410-77-0222). I thank A. Seidenberg for his helpful comments prompted by an earlier version of this article.

2. Numerical Representations in North American Rock Art

William Breen Murray

Where and how did numerals first originate? At what time in man's prehistory do these symbols of his cultural consciousness first appear? These are questions which tantalize both the mathematician and the prehistorian, each of whom is curious to know more about this important event. Yet, until recently, archaeology has not been able to provide much material evidence bearing on the early forms of numbers. Some progress has been made (see, for example, Marshack 1972), but questions such as these have gone unanswered to a large degree.

It is the purpose of this study to examine a new source of archaeological data, prehistoric rock art, which may shed light on the origins of numerals. This material is of two principal types according to its mode of execution: petroglyphs carved or pecked into the native rock, and pictographs which employ natural pigments applied over a rock surface. For our purposes, these prehistoric creations are considered simply as ancient symbols of which only a small percentage are unmistakeably representational. Given present technical capabilities, there is no way to date rock art in absolute terms and it is difficult to demonstrate any direct cultural association with other remains. As a result, rock art has traditionally been considered uninterpretable by most archaeologists.

While rock art is not found everywhere, it has been found somewhere on every continent inhabited by man, and in certain areas it forms an important part of the archaeological

record. Its occurrence depends on the availability of suitable rock surfaces in the natural environment and the presence
of some cultural tradition which sustained its production.
In some cases, as in Australian aboriginal rock painting
(Mountford 1964), rock art is substantially contemporary and
can be studied by ethnographic analogy in considerable
detail. More often, rock art cannot be linked to present-day
cultures. Rather, as with the early cave art of Eurasia, it
seems to have archaic origins in the Paleolithic hunting cultures of the last Ice Age. Each tradition of rock art production must be examined independently using a minimum of
ethnographic assumptions, for rock art may sometimes be very
old and may relate to primitive hunting adaptations in pre-
modern environmental landscapes. In the northern Mexican
periphery of the Great Basin rock art is abundant and varied,
and almost certainly seems to be a trait of considerable
antiquity (Murray 1979a).

It is maintained here that some petroglyphs and pictographs found in northern Mexico represent early numeration
systems. We assume that a particular example of rock art is
more likely to be a number if its graphic representation displays certain logical properties of the numbering process,
such as symbol repetition, graphic symmetry, and complex
ordering. With the aid of these logical tests, potential
numerical representations may be identified by discovering
and exhibiting their systemic properties independent of any
specific cultural context.

TALLIES AND DOT CONFIGURATIONS IN NORTH MEXICAN ROCK ART

The scene of our field explorations lies some 300 km north
of the Mesoamerican "frontier" in the semi-arid basin-and-
range country of northeastern Mexico, not far from the modern

industrial center of Monterrey. In this region, extensions
of the Eastern Sierra Madre form an escarpment which divides
the lowland Gulf coastal plain of eastern Nuevo Leon and
Tamaulipas from a higher, more desertic interior plateau in
western Nuevo Leon and Coahuila which forms part of the Great
Basin. The interaction of varied elevations and orographic
rainfall patterns creates a complex environmental mosaic,
from high desert in the major rainfall shadows, to pine forests on the mountain heights, and lush sub-tropical vegeta-
tion in the protected and well-watered lower canyons and
flood plains. Human occupation has been identified in this
canyon transitional zone for the past 10,000 years (Nance
1971), but rock art is abundant in the desertic interior,
where low denuded rock ridges often dominate the landscape.
Here over thirty rock art sites have been located in five
years' informal exploration, and there is every reason to
believe that many more await discovery. It is one of these
desert sites which stimulated our interest in the question of
petroglyphic numbers.

Presa de La Mula lies in the center of a small cup-shaped
interior drainage in the high desert zone near the Nuevo Leon
- Coahuila border. While rainfall is exceedingly scarce, it
is sufficient to charge the inclined strata and maintain a
small but fairly constant pond-wetland at the lowest point of
the 10-20 km wide cup. A small modern ranching community of
about 100 persons depends on this water today, and is located
about 1 km from the pond. Between the two is a low ridge of
exposed limestone, and on its west and south face about 500-
1000 prehistoric petroglyphs have been preserved on the crest
and on scattered fallen rocks. One of these, near the southern tip of the ridge is unique.

The petroglyph occupies about 3/4 of a block of the crest

Fig. 2.1. A petroglyphic count stone at Presa de La Mula, Nuevo Leon.

face which measures about 2 m across and 1.5 m high (fig. 2.1). A number of different motifs (some super-imposed) are recorded on this block, but the largest and apparently most recent is a tally recorded within a complex grid of six horizontal lines and four vertical sections. In all, 206 tally marks are recorded within its 24 component cells according to our best reconstruction. Symmetries in the counting pattern generated by the grid suggested immediately that number was an intentional property. Identical total numbers were identified on two pairs of horizontal lines, and suggested that the tallies were recorded and should be read in that direction. Moreover, a number of cells recorded the same numbers, sometimes in vertical or horizontal succession. Rows or columns of tally marks are an ubiquitous motif in North American rock art, and some (see, for example, Kirkland and Newcomb 1967) have suspected that they may be numbers. Others (see, for example, Heizer and Hester 1978) have given these petroglyphs quite different interpretations, however, and cast doubt on whether their numerical properties are intentional. In this particular example, the combination of the tally marks and the complex grid make a numerical interpretation of the petroglyph almost inevitable.

For all its regularities, however, the count stone's cultural context remained mysterious. The numbers of the grid counting pattern did not relate to any obvious astronomical cycles, and in fact none of the numerical regularities among the cells could be reduced to perfect consistency. It seemed that something was being counted which did not demand such perfect symmetry, and our first hypothesis was that the count was demographic, recording either human population or perhaps animals taken in the hunt. These were the conclusions of an earlier study (Murray 1979b), and seemed the only ones

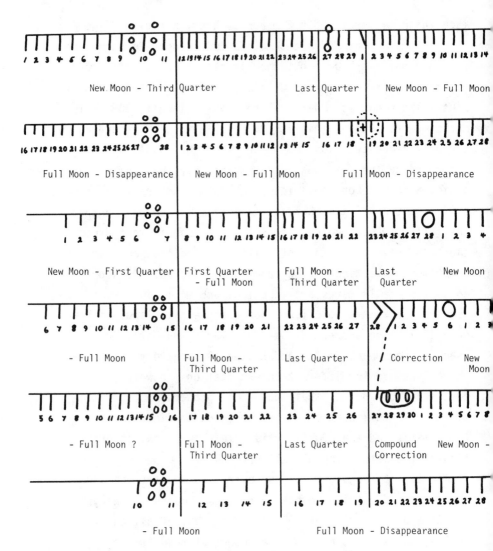

Fig. 2.2. Lunar correlation of the petroglyphic count stone in Fig. 2.1.

consistent with a hunter-gatherer band society of the type
described for the region in prehistory. More detailed study
of the glyphic evidence and additional evidence from other
nearby sites now force us to modify these earlier conclusions
considerably and to look in an entirely different direction
for their likely cultural context.

Anthony F. Aveni (personal communication) first called our
attention to the fact that the tally sum, 206, was almost
equivalent to seven lunar months, and suggested that the
petroglyph might be a record of lunar observations. This
clue led to the subsequent isolation of two additional notations which we now believe to be related to the counting pattern. One of these is a lobular glyph found in two slightly
variant forms on horizontal lines 1, 3, and 4 of the count
which we interpret to mean "completion". The other glyph we
understand as introducing a "correction", and appears on
lines 4 and 5 as an arrow-shaped extension of the tally mark
itself. When these supplementary glyphs are taken into
account, a new counting pattern is generated (fig. 2.2) which
we believe can be related in its entirety to observable portions of the lunar cycle. This lunar correlation is, we
believe, the only one consistent with all the graphic
evidence.

The lunar count begins on line 1 with 22 days observation
from New Moon to Third Quarter. The first vertical grid line
is then carved, dividing the stone in half, and cutting the
first horizontal line into two numerically equal portions
(the only line which shows this characteristic). The next
section includes four tallies, followed by an abbreviated
vertical orientation line (not counted), and a proposed completion glyph at the number 27, a respectable approximation
to an observable lunation. Two tallies follow which we take

to represent the nights of conjunction, and the final section
of line 1 counts 15 days from New Moon to Full Moon, and
begins the second month of observations. The count of the
second month is completed on line 2 section A, and its initiation and termination generate the other two vertical lines
of the grid. According to our analysis a 28-day month is
recorded, and no conjunction period is indicated after the
second month. The third month, which occupies the rest of
line 2, counts a 29-day month divided into segments of 15
(12 + 3; New Moon - Full Moon?) and 14 (3 + 11; Full Moon -
Disappearance?). On line 3 the counting pattern for the
fourth month shifts once again, and this time the four quarters are recorded with a counted value of 7 each. Complete
lunation at 28 days is marked by the proposed completion
glyph. The rest of line 3 initiates the fifth month, and,
from this point on, the grid lines are no longer the sole
determinant of the lunar counting pattern. The fifth month
is counted from New Moon to Full Moon on lines 3 and 4 (5 +
10) and the Third and Last Quarters are recorded in sections
B and C of line 4 to reach a total of 27. At that point our
reconstruction indicates that 142 days would have been
counted (29 + 28 + 29 + 28 + 28 = 142), whereas observation
would show the true value of five lunar months to be 147 or
148. We feel that the presumed correction glyph which follows introduces the compensation for this six-day error, and
brings the count in line with observation once again. The
next month's observations which finish line 4 count New Moon
to Full Moon as 16 days, and reach the same number 27 as
recorded in the first month. Then follows a unique correction glyph which is really an extension from the same glyph
on the line above, and counts three. Could the objective of
the entire exercise have been a calculation to arrive at a

30-day lunar month? We really can't be sure. The remainder
of line 5 and the abbreviated line 6 count a seventh month in
segments of 1-15 and 16-28. At this point, observation and
counting would have been less than one-half day out of phase,
and our lunar observer apparently felt his task completed.
Or did he simply run out of rock?

If the lunar correlation is accepted, two broad characteristics of lunar observation can be inferred from the
petroglyphic counting pattern. First, at least three ways of
recording the lunar month are represented, and between them
make use of all observable points in the lunar cycle. The
first divides it into two unequal portions (New Moon - Third
Quarter - Disappearance), the second into equal halves (New
Moon - Full Moon - Disappearance), and the last into four
quarters. The longer regularities are thus built upon the
observation of shorter cycles of varying lengths. Secondly,
the presumed correction made on line 4 is introduced at day
142 to reach day 148, and corresponds exactly to five lunar
months.[1]

The La Mula count stone also has an irregular "drip line"
of dots which crosses all six horizontal tally lines. Aveni
has noted the possible use of dot motifs to record the 260-
day ritual calendar at Teotihuacan and other important Mesoamerican centers (Aveni et al 1978; Aveni 1980). Folan and
Ruiz Perez (1980) have traced the dissemination and transfor-
mation of the motif up to the northern Mesoamerican frontier
and the U.S. Southwest. Based on these broad similarities
Aveni proposed that the dots on the La Mula count stone might
also have numerical significance. If the total number of
dots is added to the tally, the sum is 259 according to our
best reconstruction, only one number off from the 260-day

calendar. The similarity is enticing, but cannot be affirmed conclusively on the basis of this one example. Fortunately, another petroglyph at a site not far away makes it perfectly clear that the dot was also used to register numbers.

About forty km to the east, at Boca de Potrerillos is another major concentration of rock art. Here approximately 3000 petroglyphs are found along two km of a ridge crest on both sides of an arroyo which cuts through it at this point. This same arroyo drains from the Presa de La Mula basin, and a relationship between the two sites could easily be postulated even if glyphic evidence to this effect were totally absent. As it is, glyphic similarities are abundant, and particularly so in the case of a unique dot configuration on the North Crest just above the canyon mouth (fig. 2.3). This petroglyph is carved in a protected bay of the ridge crest which affords a panoramic view to the south and east, and is accompanied by a number of other figures which affords a panoramic view to the south and east, and is accompanied by a number of other figures which may form a composition. The central motif is a triple arc of dots arranged in an elongated loop (fig. 2.4). To these have been added some larger and smaller dots in the middle of the loop, an arching line which closes the figure at the top, and a long curved "handle" to the right. The overall shape looks something like a jar.

The counting pattern of the Boca de Potrerillos jar is very different from that of the La Mula tally count, and no explanation of its internal divisions can be offered at this time. On closer inspection each arc of dots seems to be continuous; none manifest any prominent internal sub-dividing mark, nor do the arcs count side by side. The sequence of production of the arcs cannot be determined, nor can we say

Fig. 2.3. A petroglyphic count stone at Boca de Portrerillos, Nuevo Leon.

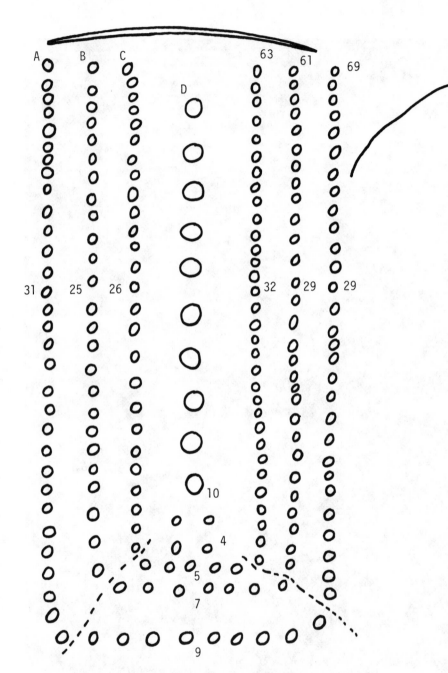

Fig. 2.4. Schematic drawing of the count petroglyph in Fig. 2.3.

whether they were cut from left to right or the reverse. A
viable working hypothesis would be that the inner arc was
carved first, since it is the least crowded, but this may be
quite accidental. Only the total number of dots in each arc
can be determined with any degree of confidence, and even
this is subject to some doubt (± 2) due to the broken down
condition of the dots at the upper right of the outer two
arcs. Our best reconstruction identifies 63 in the inner
arc, 61 in the middle, and 69 in the outer totaling 193 dots,
none of which are numbers which can be related either to the
La Mula tally count or lunar cycles. The total is nonetheless a significant sum when we take into account the dots
enclosed within the loop. Ten larger dots are carved inside
the jar with a square of four smaller dots at the very bottom, and give every evidence of being part of the same motif.
When they are summed to the arc dots, we reach a total number
of 207, very close to the La Mula tally, and an underlying
similarity between the two which can hardly be fortuitous.
We conclude from this that the dots must truly be numerical
in this example, and constitute another petroglyphic counting
motif just like the tally marks.

The solution to the Boca counting pattern may well lie in
more complex relationships with other glyphs around it. One
in particular, an approximately symmetrical dot grid just to
the right of the jar (fig. 2.5) shows the same patina, and
seems especially likely to be related. The total number of
dots is 69, and repeats the sum of the outer arc of the jar
count. If this is summed to the dot arcs (193 + 69 = 262) a
number is reached which is close to the 260-day ritual cycle.
But the validity of the operation remains tenuous, for the
sum of the total counts (207 + 69 = 276) strays much further
from the ritual mark. We take the dot grid to be a separate

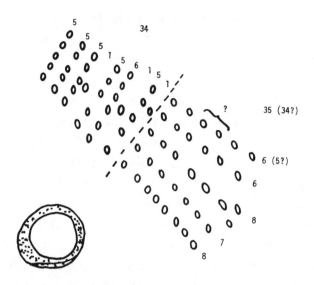

Fig. 2.5. A dot configuration petroglyph on the North Crest adjoining the count petroglyph in Fig. 2.3.

count from the jar glyph but recorded in the same system. In this case its principal importance is in revealing possible component numbers within one of the arc totals. Looking closely, for example, we see that the dot grid can be divided into two components (34 and 35, respectively) which were apparently cut with different orientations. The upper seems to run in more regular horizontal lines, while the lower appears to be oriented on the vertical and is not completely symmetrical. The component lines of both grids show closer correlation to observable lunar periods than either of the sums would indicate. The upper component can, in fact, be broken down into a 28-day lunar month with a 6-day "post-fix" using counting combinations found in the La Mula tally count. But a complete lunar correlation is still not apparent.

Other glyphs besides the dot grid may also help to clear up the question. To the left of the jar are two splayed

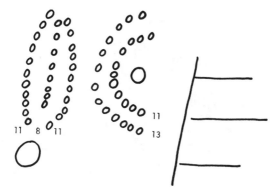

Fig. 2.6. A dot configuration petroglyph on the South Crest adjoining the count petroglyph in Fig. 2.3.

vertical lines of large dots (7 + 7 = 14), while above to the right are another pair of large dot arcs (9? + 9? = 18?). Below are two horizon lines, a circle, and more large incised dots. Do they represent other astronomical events or are their numbers to be summed in some combination? The evidence favoring either alternative is still inconclusive, and only additional evidence of dot counts can lead us closer to the solution. Fortunately, at Boca such dot configurations abound, and open up various comparative possibilities.

Just across from the jar count on the South Crest's lower spine, for example, are two other dot configurations in close association with circles and possible horizon lines (fig. 2.6). The numbers of the one total 30, and its three component dot lines are symmetrical. The curved outer arcs repeat the 11 + 11 sequence which initiates the La Mula tally count. The other configuration is a double arc (11 + 13 = 24) flanked by three vertical and one horizontal line, which may be related to the whole figure. Once again, we cannot be sure whether the two figures refer to the same event or different ones, and whether the center circles are part of the

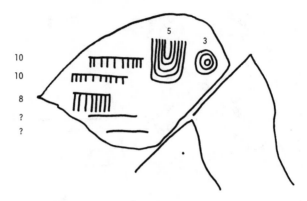

Fig. 2.7 A possible count petroglyph at Paredon, Coahuila.

counting pattern or have a strictly non-numerical symbolic
function. If all the dots are summed (30 + 1 + 24 + 1 = 56)
to the three horizon lines, we reach a number (59) which is a
very close approximation to two lunar months, but we have no
way of knowing whether this procedure is justified.

What seems clear is that naked-eye astronomy may be somehow related to the petroglyphs at Boca de Potrerillos. Evidence presented elsewhere (Murray 1982) suggests that part of
the site may have functioned as a base point for reading
horizon alignments at the solar equinox and solstices.
Indeed, it is possible that many different kinds of horizon
rising and setting alignments are encoded, and there will be
no simple solution to the inter-relationships between numbers
and astronomy. But we propose that they are related in some
still-to-be-determined way and that the petroglyphs are the
mediating symbols. Another site near Presa de La Mula is
ideally situated to function as a good astronomical observatory. This is a large rock promontory overlooking the bend
of a large dry arroyo near Paredon, Coahuila, only about 15
km from La Mula. This rock commands a striking view of the

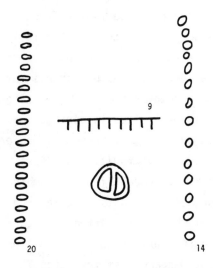

Fig. 2.8. A dot configuration petroglyph at Presa de La Mula, Nuevo Leon.

surrounding landscape, and the petroglyphs carved on it (fig. 2.7) include a triple concentric circle and three tally lines (10 + 10 + 8 = 28). Between the two figures is a quintuple inverted arc similar to the shape of the jar count at Boca. Here the lines are solid and cut deeply into the rock. The three figures seem to be interrelated, but the partial erasure of a section of tally counts just below makes an exact reconstruction of the numbering pattern a bit risky. The evidence does suggest, however, that the U-shaped glyph may also function as a cycle counter in some contexts. The possible use of the promontory as an astronomical observation point is hinted strongly by the combination of the physical setting and the associated glyphs.

Another petroglyph not far from the count stone at La Mula (fig. 2.8) illustrates a possible combination of two types of dots with a tally line and a halved circle. The counting pattern (20 + 9 + 14 = 43) can be correlated closely to 1.5

lunar months, and is very close to the first line of the
count stone. It shows that both dots and tally marks may be
combined to register numbers, and brings our initial tour of
the petroglyphic evidence for number counts full circle.[2]

Using the La Mula count stone as our paradigm, we have
noted evidence at several sites in northeastern Mexico which
may link count petroglyphs to astronomical observations.
This petroglyphic counting tradition utilizes distinct counting symbols, either independently or in combination, which
may register chronological counts ranging from 15 days up to
seven months. From these observations the question naturally
arises: Is the petroglyphic counting tradition a reflection
among barbarians of Mesoamerican traditions? Or is it rather
an Archaic tradition of considerably wider distribution out
of which Mesoamerican counting developed? The first alternative accentuates diffusion, while the second stresses evolution, and the choice between them can not be made with the
evidence presently available. Our present inclination is
toward an evolutionary explanation, and the reasons are
locked within other petroglyphs at the same sites we have
been discussing which make use of the dot and tally mark in
non-astronomical and non-numerical ways. These negative
examples, we feel, constitute important evidence in defining
the origins of numbering conventions.

NON-NUMERICAL GEOMETRY AND SYMBOLIC EVOLUTION

We begin this examination with a dot configuration which
reappears in three different forms, and may well be a number.
If so, it would be three ways to write the number 32. This
number appears to have no special lunar or astronomical significance, but it must have been important enough in some
other context to bear repeating.

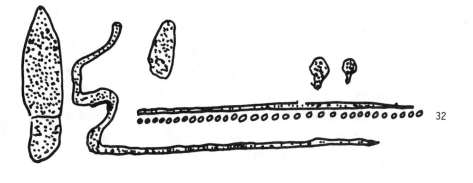

a

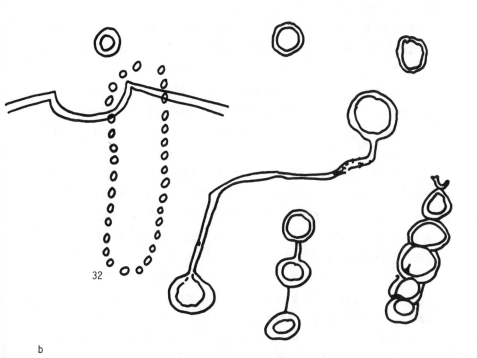

b

Fig. 2.9. a-b. Dot configuration petroglyphs at Paredon, Coahuila.

The glyphs which accompany the 32 dots provide conflicting
clues about this missing cultural context. The first two
examples come from a site near Paredon, Coahuila which is
dominated by carvings of oversize projectile points and other
stone (?) objects in a very distinctive incised relief style.
This same style is present at Presa de La Mula, and the
point-shaped figures beside the tally count are typical examples. One of these projectile points also flanks a complex
figure (fig. 2.9a) within which 32 dots are registered just
below a deeply-incised orienting line. The associated glyphs
include a serpentine shape just in front and below the dot
row, a rectangular oblong shape (a scraper blade?) just
above, and two "mushrooms" above the orienting line. The
context of the "counting" here seems related in some way to
hunting. The other example at Paredon (fig. 2.9b) contains
no such references in the associated glyphs, and leaves us
perplexed again. It is a dot loop superimposed among circles and wavy lines which may or may not be part of a single
composition, and could mean almost anything. The third example is more evidently numerical (fig. 2.10), for here the
total arc of 32 dots has been divided at the mid-point (16)
of the count by the addition of an extra dot. This petroglyph, found at Narihua, Coahuila some 100 km further inland,
thus displays both symbol repetition and ordering, and suggests that the number is no accident. But the accompanying
glyph, a lightning bolt zig-zag, doesn't tell us very much
about the symbol's context, and is dissimilar to the contexts
at Paredon. We can only conclude that dots were used to
record other things than astronomical observations in the
petroglyphic counting tradition.

Many other examples of petroglyphic dots are carved in
asymmetric relations, and we suspect that their intended

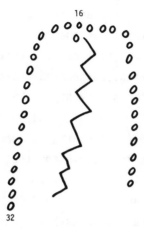

Fig. 2.10. A dot configuration petroglyph at Narihua, Coahuila.

meaning is completely non-numerical. These dot configurations fall into three basic types. One kind are long sinuous
lines of dots which follow natural contours of the rock,
sometimes for several meters. One interesting example of
this type is found in the smoke-blackened roof of Cueva Ahumada, where petroglyphic dots have been painted white, and
follow a prominent crack in the roof into the rock-choked
depths of the cave. A similar line crosses exposed bedrock
on the open slope nearby. At Boca de Potrerillos, Presa de
La Mula, and other sites the sinuous lines cross vertical
rock faces, and look like "drip" lines. One such "drip"
crosses the tally count. Another type of dot configuration
appears to be irregularly spaced within enclosing forms, or
the dots themselves create the form. A notable example of
this latter type is a half-circle of dots pecked on the edge
of a broken rock in Icamole canyon, near Cueva Ahumada, with
three "halo" bands of pecked dots arching above it. Here the
number of dots seems to be totally subordinate to the

rendering of the form. A final type consists of dot fields
or grids which are plainly asymmetrical. Although some of
these may be numbers we still can't detect, it appears that
many others may represent abstracted design elements such as
beadchains, or even the starry night sky. Examples of this
type are found at all of the sites we have discussed. All
three of these types of dot configurations delineate forms or
textures, and lead to the conclusion that some petroglyphic
dots are non-numerical.

Tally "counting" can also be found in similarly ambiguous
contexts. One such case is a stylized deer antler with the
horns arranged in a row of 28 projecting above. This petroglyph is found less than 100 m from the tally count at La
Mula. Similar antler shapes are known from several other
petroglyph sites. At Boca de Potrerillos we even have an
anthropomorphic incised head sporting an antler-headdress
similar to those found preserved at Cueva de La Candelaria,
near Torreon, Coahuila, in burial contexts (Aveleyra et al
1956, p.117). Only the La Mula example, however, reshapes
the horns to form a "count". While the number 28 might be
construed as a lunar cycle, it appears that an interpretation
as "animals taken" is more economical and consistent with the
other known examples. Tallies may also be numbers in contexts other than lunar or astronomical counting, and must be
analyzed case by case.

What do non-numerical or doubtfully-numerical petroglyphs
tell us about the more limited group we have associated with
the petroglyphic counting tradition? Do they lead us any
closer to the identity of the petroglyphic counters? Here we
believe the evidence points basically in one direction, without resolving whether evolution or diffusion is involved.
Only the symbols of a hunting and gathering culture accompany

the many examples of counting symbols we have encountered in northern Mexico. Lacking firm chronological control, we can not tell whether the relationship is ancestral or parallel; it could even be both. But with the help of the above non-numerical examples, we can develop a speculative scenario of how one type could have evolved from the other in response to the conditions and necessities of a desert hunting-gathering culture.

The evolutionary scenario might begin at some still undetermined early post-glacial time horizon. Northeastern Mexico is occupied by peoples practicing an Archaic desert lifestyle among whose traits we can identify petroglyph-making, and very probably an intense interest in the sky. There is good reason to believe that astronomical knowledge would have been specially prized, for it confers a tremendous adaptive advantage among nomadic desert peoples. It permits more precise scheduling of movements between varied food resources, many of which (such as migratory birds, for example) might be available only for short periods. Sky-knowledge becomes the key to land navigation, and he who moves fastest and most efficiently eats best.

What seems clear is that the first graphic use of circles is representational. The symbol's meaning is ambiguous, and the range of possible depictions very large. But from the arrangements of the petroglyphs it is clear that none show evidence of complex ordering: all are in freer asymmetrical patterns.

At some later date more complex geometrical forms begin to appear in conjunction with styles which are more evidently representational. The geometrical forms are of two principal types: grids of dots, and linear rectangles. Both of these motifs show properties of formal order, and must be regarded

as probably abstract and non-representational. Although
their numerical patterns have not yet been fully studied, the
La Mula tally count suggests that the fusion of the two was
basic to the creation of the petroglyphic counting tradition.
It is also likely that both were used independently as counting symbols, and a wide variety of other symbols may also
have been used, perhaps even different ones for each context
in which the need to enumerate arose. Our working hypothesis
is that at a very early point the dot came to represent a
day, and then came to be used schematically to represent pro-
gressively longer sequences of days. The counting sequence
is superimposed over a geometrical form of given symbolic
value (as in the Mesoamerican examples found by Aveni and
Folan), and thereby fuses numbering to a context.

   The tally mark, on the other hand, seems to be associated
more directly with hunting representations of different
kinds. The antler count at La Mula is decisive evidence in
this regard, even though it does not tell us how petroglyphic
counting and hunting were linked. Heizer and Hester's
hypothesis that tallies are fence representations may well be
correct in some examples, and if so, we can identify a representational source for what later became a number symbol just
as with the dot. But whatever the representational source,
we can still imagine many things the deer hunter might want
to count. Times of migratory movements, days of rut, or the
numbers of animals taken might all be important numbers in
either a ceremonial or a utilitarian sense. From one or
another of these hunting contexts, the tally may have developed into a counting symbol too, and completed the basic sym-
bology of the petroglyphic counting tradition.

   The existence of other counting symbols within the

tradition seems likely, but not fully demonstrated. We have mentioned some of the evidence, but far more remains to be analyzed. The use of cycle markers, for example, is possible but not proven. At no point, however, do the petroglyphic counts follow the symbolic conventions of the Mesoamerican counting system . Each symbol is used independently, and when they are combined, they are not combined in the Meso-american way. Petroglyphic counts are limited to simple binomial combinations. Small numbers are typically repre-sented, much as we might expect among hunting-gathering nomads. A monomial symbol is quite adequate for recording these small amounts, but a system of monomial tally counting must have some logical outer limit beyond which it tends to break down into binomial combinations. Taken together, the La Mula tally count (binomial) and the jar count at Boca de Potrerillos (monosymbolic) seem to mark this limit rather clearly at around 200. The vast majority of tally and dot counts must register numbers considerably smaller. We can only assume that these shorter periods and smaller quantities must be more typical of the hunter-gatherer's perception of his environment than the billions on Wall Street which dance before modern man's eyes.

NOTES

1. Recently located petroglyphic dot counts at nearby Icamole, Nuevo Leon (Murray 1985) make it clear that lunar eclipse intervals are being recorded. The La Mula count stone registers the 148-day and 177-day intervals, while Icamole Petroglyph 2-III records the 162-day interval (see Aveni 1980, p.76 and Table 4, for the alternative intervals).

2. A similar form of lunar counting has also been found on Late Prehistoric bone artifacts from a burial near Corpus Christi, Texas (Murray 1984) and may have been portable versions of the petroglyphic counts.

ACKNOWLEDGEMENTS

The author expresses thanks to the following persons for their contributions to this study: Dr. Anthony Aveni of Colgate University for his many valuable suggestions on possible astronomical correlations of the rock art described here; Drs. Thomas Hester and Joel Gunn of the University of Texas - San Antonio for their helpful comments and criticism of the interpretations expressed; and to the staff members of the Centro Cultural Alfa, students of the Universidad de Monterrey and the Escuela de Antropología, Universidad Autónoma de Nuevo Leon, for their fellowship and enthusiasm on the numerous field visits required for these investigations.

3. Some Notes on Quantification and Numerals in an Amazon Indian Language

Maurizio Covaz Gnerre

The intellectual activity known as "mathematics" was developed in certain areas of the world in which a tradition of writing had already been established. However, most cultures did not, and still do not have such a tradition. If we state that the mathematical skills or the systems of numeration and calculus in these cultures are rudimentary, we make a judgement from the wrong perspective. Indeed, we use the exception -- cultures with a written tradition, especially a western intellectual tradition -- as a measure for all others. If we do not base our judgement on traditional western standards, can we still look for mathematical skills in cultures without a written tradition? In my opinion we can, but we have to redefine the object of our search. What should we look for and which concepts should be considered relevant? Should we base our search on an operative comprehension of the concept of number? Or, should we consider most relevant an ability to establish one-to-one relations between objects, fingers, stones, or shells?[1]

A basic hypothesis of this paper is that there is internal variation in all human groups and cultures. In cultures where we cannot observe "mathematics" as we traditionally define it, we should look for basic capacities on which a mathematical intellectual activity can be built. However, as some degree of variation is always present, we should look for both the presence of an operative comprehension of numerical concepts as well as a use of number which does not imply such a comprehension.

It seems to me that the postulate of variability is necessary. A uniformitarian perspective of so-called "primitive"
societies, so often implicit in the works of anthropologists
and linguists, prevents us from understanding many aspects of
those societies and encourages us to view them as static
entities. In reality, we find some degree of individual
variability in skills and intellectual achievements in even
the smallest human groups. We find a dynamism that is
related to the existence of variability. While this perspective might be obvious for many readers, it is not very widespread. For example, one encounters statements such as "the
people X do or say such and such" even in the most recent
anthropological literature. Nevertheless, in anthropological
tradition the assumption of variation in intellectual skills
is implicit. Some persons are more reflexive, more knowl-
edgeable, or more curious than others. These are the infor-
mants each anthropologist in the field would like to find as
a reliable associate in his or her research. To attribute
the same capacities of reflection to all the members of a
community is an unfortunate oversimplification. A uniformitarian and static perception of "primitive" cultures is an
inadequate one. It is my claim that the most significant
perspective in approaching the problem of mathematical skills
in cultures without a tradition of literacy is one based on
synchronic variability and dynamism.

In most cultures without a written tradition a capacity
for abstract reflection is achieved by only a few individuals
and cannot be accumulated and passed on from one generation
to the other. The reflection of a single Thales starts and
dies in his or her lifetime, without a chance of spreading or
of being fixed for future reflection. Each Thales begins
again on the base of his or her own culture.

Recently the Piagetian anthropologist C.R. Hallpike (1979, p.62) wrote:

"it seems insufficient to argue that wherever there is a demand for a cognitive skill to become general in a society, that demand will be satisfied. [...] Merely because a society manages to get by with its existing repertoire of collective cognitive skills it does not demonstrate that these are wholly adequate to its needs, or that cognitive demand will be met by cognitive supply.

The fact that a society does not display a particular cognitive skill in a high proportion of its population is therefore likely to be the result of two factors - that it is not developed in many of them as a result of general environmental factors, and that it is not relevant in that society, though as we have just noted, 'relevance' is not at all easy to define."

Although I agree with the general view expressed in these statements, I would argue that a 'demand' for cognitive skills can become widespread in a society and some members of it can satisfy that demand in some way even if their intellectual achievement does not become general and is not preserved for other generations. Without a written tradition, innovative knowledge and reflection can hardly become a seed for future elaboration.

Yet, spoken language can be a very indirect 'register' by which some concepts find their way into a cultural tradition. This is a slow process of lexical change and innovation. However, with a perspective based on variation and dynamism we cannot claim that language is a mirror of the shared and unshared cognitive representations of its speakers. The use that many western anthropologists and linguists have made, and still make, of linguistic data is a consequence of a

uniformitarian and static view of 'primitive' cultures. Many
of them have considered language as a primary indicator able
to reveal the degree to which the speakers could or could not
deal with mathematical concepts. An idea still alive today
is that limited skills or lack of interest in quantification
and numbers are represented by a very limited numerical
terminology.

On the other hand, a somewhat different and very engaging
position can be found in the research of Tylor and Levy-
Bruhl, concerning numeration in 'primitive' societies. Tylor
(1903 [1871], vol.I, pp.242, 246), in his chapter on "the art
of counting", noted that:

"Among the lowest living men, the savages of the South
American forests and the deserts of Australia, 5 is actually
found to be a number which the languages of some tribes do
not know by a special word."

But, the same author went on to say:

"men counted upon their fingers before they found words
for the numbers they thus expressed [...] in this department
of culture, Word-language not only followed Gesture-language,
but actually grew out of it."

Levy-Bruhl (1926 [1912], p.205), some forty years later,
also stated that the existence of counting skills could be
possible without number names in a language and claimed that:

"It is a mistake to picture the human mind making numbers
for itself in order to count, for on the contrary men first
of all counted, with much effect and toil, before they conceived the numbers as such."

This position appears to represent a historical perspective but was not extended to a reflection of synchronic data.
It represented, at least for Tylor (1903 [1871], vol.I,
p.251), an important step toward the historical understanding

of the origin of numerical terminology. Tyler wrote:

"The theory that man's primitive mode of counting was palpable reckoning on his hands and the proof that many numerals in present use are actually derived from such a state of things, is a great step towards discovering the origin of numerals in general."

From a synchronic perspective a limited set of numerals in a language (and in a culture) does not imply that all the members of that group can do only what the lexicon permits. At least some members of the group can count and calculate beyond the 'limits' of the language. On the other hand, I agree with Hallpike (1979, p.245) when he writes:

"even where a culture possesses an extensive verbal system of numbers, we are not entitled, from the fact alone, to deduce that the members of that culture have an operational grasp of number. Such a conclusion must rest on empirical evidence of the way in which their numbers are used."

The problem of expressing numbers and quantities is a challenging one for a conservative view which considers language as merely a **verbal** system of communication, because the use of parts of the body, mainly the hands, is very clearly associated with verbal expression. Both Tylor (1871) and Levy-Bruhl (1912) pointed this out when they referred to the widespread use of signals which in a systematic way integrate the verbal numeration system in many languages and cultures. In many cases body movement and hand movement play a role in communication which is far more relevant than is generally recognized. Gestuality is present not only in 'contextual' or 'deictic' communication but also in discussion of highly abstract topics.[2] In many cultures the use of fingers, toes or other parts of the body is an essential component of numeration. In some groups of Melanesia we find numeration

systems which involve the use of the whole upper part of the
body to count up to 47 (Hallpike 1979, pp.240-241). This
systematic use of the body in numeration seems to me worthy
of research and reflection. Because the use of hands or body
signals is relevant in numeration, many linguists and anthropologists had to include the description of non-verbal systems which, together with words or utterances, were used to
express numerals.

With the foregoing ideas in mind, I will discuss the case
of a culture and a language which has been considered from a
traditional perspective without any real appreciation of the
mathematical skills manifested by its members. I will analyze some aspects of the language and culture of a native
society of the Upper Amazon, the Shuar of Ecuador.[3] By
observing the available data we gain insight into the slow
growth of a set of numerals and, consequently, into the dynamic process of the language and of the culture.

A good account of the way in which the Shuar used to count
can be found in the first ethnographic monograph on their
culture (Karsten 1935, pp.548-549) written on the base of
field data collected between 1916 and 1917, and between 1928
and 1929. I will quote the somewhat lengthy description
because it illustrates in a very clear way the complementa-
rity of verbal and non-verbal communication.

[Quotation begins]
The majority of the Jibaros are able to count to 'ten',
but only for the five first numerals have they proper names.
They always count with the fingers, beginning with those of
the left hand, and then also with the toes.
Among the Upano tribes the numerals are as follows:

chikíchi, one

hĩmera, two

menéindu or kámbatama, three

eínduk-eínduk, four

wéhe amúkei (= "I have finished the hand"), five.

If the Jibaro Indian is obliged to count more than five, he seizes the fingers of his right hand with those of the left; then, beginning with the thumb, he goes on counting, using the following expressions:

> huíni wéhe, six, (here I have a [finger from the other] hand)
>
> hĩmera wéhe, seven (two fingers [from the other hand])
>
> menéindu wéhe, eight (three fingers)
>
> eínduk-eínduk wéhe, nine (four fingers)
>
> mai wéhe amúkahei, ten ("I have finished both hands").

If it is necessary to continue counting the Jibaro seizes the toes of one foot, one by one, and counts chikichi, hĩmera, menéindu (one, two, three), etc. When he arrives at the fifth toe he says: huini náwi amúkahei, "here I have finished one foot" (Huíni = here, náwi = foot), it being understood that he has begun with the hands. The said phrase, therefore, is equal to fifteen.

Thereupon he may continue counting with the toes of the other foot, and when he arrives at the last toe he says: mai náwi amúkahei, "I have finished both feet"; this means 'twenty'. Twenty is the absolute limit for the Jibaro counting, as far as it is expressed in words.

'Ten' may also be expressed by joining both hands closed, without using any particular expression, and if the Jibaro wants to indicate 'twenty', 'thirty', 'forty', etc., he does

it by joining his closed hands two, three, four, etc. times.
It is characteristic of the Jibaro Indian that he cannot
indicate a number in the abstract, but always does it by
signs, even when he possesses a particular word for it, and
nearly always he begins from 'one', counting on his fingers
until he arrives at the number he wants to indicate.

[Quotation ends]

In one early grammar of the Shuar language (Ghinassi 1938)
written by a missionary who spent many years among the Shuar,
we find a description which is very similar to Karsten's.

"1) After twenty the Jíbaro Indian does not have any
other number but he can count as much as he wants repeating
with the closed hands the value for 'ten' and adding with the
fingers the unities he needs. Between one value for ten and
the other he says the adverb - atáksha (and again) - or júsha
- (and this).

2) Ordinarily the Jíbaro Indian does not count more than
fifteen or twenty values for 'ten' because it would be difficult to remember them; so that for a bigger number he uses
the word untsúri (very much) - píshi (plenty) or a comparison
- uéka núke (as ants)." (Ghinassi 1938, p.85).

These two descriptions[4] show how non-verbal communication
plays a central role in numeration and quantification and
that quantities up to an undetermined limit are easily recognized. It is clear that the main problem for both authors is
a traditional western point of view which places emphasis on
what people can say. Both authors have to admit that the
Shuar, when they have to, are able to count beyond the limits
that would be represented through verbal expression. This
hypothesis was implicit in the above citations: "If the
Jibaro Indian is obliged to count more than five...", "If it

is necessary to continue counting...", "Thereupon he may
continue counting..." (Karsten), and "After twenty [...] the
Jibaro Indian does not have any other number, but he can
count as much as he wants..." (Ghinassi). With this perspective it is possible to edify an image of the "primitive". In
relation to numerical terminology, this perspective was
explicitly stated by Tylor (1903 [1871], vol.1, pp.243-244):

"It is not to be supposed, because a savage tribe has no
current words for numbers above 3 or 5 or so, that therefore
they cannot count beyond this. It appears that they can and
do count considerably farther, but it is by falling back on a
lower and ruder method of expression than speech - the
gesture-language."

This perspective is the same one that leads some linguists
to state that in a language "there is not" a comparative construction, but that if the speakers really have to verbalize
a comparison they can use "some other" syntactic construction
of the language. It is quite obvious that, in a case like
this, a specific syntactic model -- the comparative construction present in most of the Indo-European languages -- is
taken as an arbitrary measure for the other languages. We
will go back to this specific example when we mention some
recent developments in the Shuar mathematical terminology.
The main point here is the arbitrariness of similar judge-
ments. In the anthropological literature it is easy to find
such arbitrary interpretations which do not contribute to the
understanding of cultures. To mention a case in point, I
quote a passage from the most recent anthropological monograph on the Shuar (Harner 1972, p.30):

"In actual practice trading partners do not keep a strict
accounting of transactions. Since a variety of valuables is
exchanged by two 'friends' at one time and since the Jivaro
do not value numeration, the exchange is often slightly

uneven."

First of all, it is only if we base our judgement on explicit linguistic data (the number names in the language) that we can say that the Shuar "do not value numeration". The fact that the exchange is often "slightly uneven" is only distantly related to mathematical skills proper. Such a statement, moreover, is an example of a judgement made with a western conception of trade and exchange in mind. It is likely that in most trading systems, as among the Shuar and other Upper Amazon peoples, the participants know that one of them receives some advantage. This is a central character-istic of the trade system which establishes and preserves social relations. If this is the case, a conclusion that the Shuar "do not value numeration" is based on a concept of tra-ding which is deeply divergent from the traditional Shuar view of it.

Another aspect of the statement we are analyzing here is that of the relevance of a concept of "numeration" in trading patterns. The problem here concerns the relevance of quanti-fying skills and establishing relations between quantities as cues of operational capacities in general. The relevance of these concepts is not at all clear. Hallpike (1979, pp.98-99) writes:

"All estimates of size, or length, or height, or quantity, or duration thus immediately conjure up sensory images or associations of familiar activities and forms of behaviour, of procedures and customary modes of coordination, and do not stimulate or require quantitative analysis or dimensional abstraction. The sensations of size, duration, weight, and heat, among many others, are thus necessarily subjectified in primitive experience. Without units of measurement and quan-tification it is very difficult to separate out particular dimensions and to compare objects in terms of them alone, or to become aware of relations of compensation between different dimensions.

[...]

In a world of gourds, pots, bamboo tubes, baskets, hollowed-out tree-trunks, string bags, and sewn-up animal skins, which are used for transporting and storing things and not for measuring them, it is extraordinarily difficult accurately to perceive displacements and conservations of quantity or area."

While this may be "extraordinarily difficult" it often happens, and when western observers find such skills in a "primitive" people they are struck by such abilities. One recent example can be found in the account of the success of the Kpelle people of Liberia in evaluating the number of cups of uncooked rice which could be contained in a bowl (Gay and Cole 1967, p.8). In that case the skills of the Kpelle people became even more evident when compared to the failure of some American Peace Corp Volunteers in the same test. From this example we could conclude that the operative use of the concept of number (which we should attribute to the American Peace Corps members) can be irrelevant to the skills of practical evaluation of volumes and quantities. On the other hand, when we find some good practical skills they should not be used as evidence of mathematical abilities.

The interaction between the activities of measuring, quantifying and finding mnemonic devices for quantities can be relevant to the growth of numerical terminology. We could assume, as a working hypothesis, that words in general, and numerals and quantitative expressions in particular, are "indices to pre-linguistic cognitive schemata, according to which we organize and remember our experience" (Kay 1979, p.1).

In the traditional culture of the Shuar and of other Jivaros we find some activities in which good skills of exact
execution and measure are requested. On the other hand, it
is hard to imagine any activity in which the capacity of
counting and calculating could be particularly helpful. We
mention three activities: the construction of a blow-gun, the
digging out of a canoe, and the building of a house.

Only some of the Achuar men[5] are able to produce long
blow-guns. They are constructed from two pieces of hard wood
which have to fit together perfectly. The internal part of
the gun cannot be even minutely uneven.

Only a few of the Achuar and Shuar men are able to manage
the work of digging out a big log of selected wood to produce
a canoe. This work requires great skill in evaluating the
volume of wood needed to yield a dugout canoe that will be
perfectly balanced once it will be put in the water.

Almost all the Shuar and Achuar men are able to build
their house, with the help of other men. The Achuar build a
new house every six to eight years. The old house is left
when the new one is ready. In general many men participate
in this activity. Consequently, this is a relatively common
activity in the life of each Achuar man, who builds his own
house several times and who has to help other men many times
in his life. In this frequent and collective activity a
standard measure is used. The name of the measuring unit is
nekapek or nekapmatai and in general it is represented by a
stick. The length of the stick is based on the human body.
It is the segment from the girdle to the ground. A relevant
question is why such a standard measure is in use in one of
the most frequent and socializing activities in the Achuar
culture.

The root of the name of the measure is nekap-. This root,

from our point of view, can be glossed with a set of meanings. It is possible, however, that for an Achuar or a Shuar
this root identifies a single conceptual area which includes
meanings such as 'to show', 'to indicate' (as in hintya
nekap-,'to indicate the path'), 'to demonstrate', 'to measure' and 'to quantify'. In the recent process of implemen-
tation of a Shuar mathematical terminology[6] this root has
been used to express the concept of 'number' and 'to count'.
The same root is related to the root neka-, 'to know, to be
aware'. From this root are derived forms such as nekas,
'right, true', or followed by the negative morpheme (-ca),
nekasmianča, 'valueless, fake'. We could say that meanings
such as 'to measure' and 'to know' are related through two
varieties of a basic root. I am not claiming that these
meanings are 'the same' for a Shuar or an Achuar, but I would
like to claim that some common component exists in such meanings as 'to know, to show, to measure, to quantify'.

When we look at the few numerals of the Jivaroan languages, from a comparative and etymological point of view, we
can reach some insight into the historical growth of the set.
Although my main concern here is with the Shuar language, I
will also use data from two other strictly related Jivaroan
languages, Achuar and Aguaruna. The first four numerals in
the three languages are as follows.

	Shuar	Achuar	Aguaruna
1	čikíčik	kíčik	bakíčik
2	hímyar	hímyar	hímaŋ
3	manáintyu	kampátam	kampátum
4	áintyuk áintyuk	učínyuk učínyuk	ipáksumat

The first fact we can point out is that while the forms

for 'one' and 'two' are basically the same in the three languages, for 'three' we find two distinct forms and for 'four'
three distinct forms, one for each language. This fact can
be interpreted as revealing that the first two numerals --
common to the three languages -- are older than the third,
and that the third numeral is older than the fourth.

The etymological analysis leads us to the same conclusion
because the first two numerals are much more opaque in etymological terms than the third and the third more than the
fourth.

In the forms for 'one' we find that kĭčik of the Achuar
represents the common segment of the morpheme. I am unable,
however, to establish whether the forms with initial čí- and
ba- of Shuar and Aguaruna are expansions of the kĭčik form.
In the Shuar language we find other similar morphemes such as
išíčik, 'a few, a little', číkič or tíkič, 'other', and
čikyá-s-, 'to separate, to stay alone'. I am unable, however, to relate any of these morphemes to the form for 'one',
in a more significant way.

The forms for 'two' are exactly the same in Shuar and
Achuar. The difference in Aguaruna is due to systematic phonological correspondences between this language and the other
two. I am unable to propose any etymology for these forms.
I can only remember the existence in Shuar of the expression
himyámpramu, 'twin', which seems derived from the morpheme
for 'two'.

Things become a little more transparent with the Shuar
forms for 'three' and 'four'. The Shuar form for 'three' can
be related to a whole set of morphemes such as ména, 'left,
left side, left hand', menánt-, 'to stop, to stay at a side',
men-ká-ka-, 'to miss, to loose', men-á-k-, 'to miss the path,
to be unable to find the path', main-ŋka-, 'spoiled' (used

for manioc beer or for cooked manioc). All these morphemes have some common meaning which ranges, it seems to me, from 'to miss some pre-existent or characteristic', to 'to be in a non-central position'. An etymological speculation could lead us to propose 'uneven' as an original meaning of the form for 'three'.

It is likely that the form manáintyu arrived in Shuar in relatively recent times, to replace an older form similar to that used at present in Achuar and in Aguaruna. Karsten (1935) stated that "among the Upano[7] tribes was used for 'three' menéindu or kámbatama." At the time he collected his data some variation between the two forms should have existed in Shuar.

I am unable to associate the Achuar and Aguaruna form to other morphemes of the two languages or to any morpheme of the Shuar. We should note that the difference in the semi-final vowels can be explained as a systematic phonological correspondence between the two languages.

The Shuar form for 'four' is probably related to the morphemes aínik, aíniu, ániu, 'similar, equal, even', ain-kia-, 'to do the same', aint-ra-, 'to go together, to follow'. We could associate the root ain(t)- with a meaning 'equal, together'. Both the Shuar and the Achuar forms are reduplicated. The semantics of reduplication in Shuar is not easy to catch because it is a widely used device in the language. I would say that it has in general an intensive meaning, so that the Shuar form for 'four' could be interpreted as 'the very even' or something very similar. The Aguaruna form for 'four' is very interesting. The Jivaros used to count by putting down fingers, starting from the minor one, so that the index is the fourth finger. The index is used to paint the face with red pigment from the Bixa Orellana. The

meaning of the Aguaruna name is 'for painting'. In Shuar the
name of the finger is ipyáksuntai, 'for painting with Bixa
Orellana (ipyák) pigment'. A related form -(u)sumtai- is
used in Shuar to refer to 'nine'.

The fact that the Aguaruna name of the number is indeed
the name of the fourth finger constitues, in some sense, the
actual link between the verbal names of the numbers and the
gestual activity of reckoning on the fingers. In the short
numerical distance from 'one' to 'four' we find a great linguistic distance from the absolute etymological opaqueness of
the first two numerals to the relative transparency of some
of the five names for 'three' and 'four'. In some sense
there is also a difference in degree of arbitrariness,
because a form which is more transparent and more easy to
relate to other forms of the language is in relative terms
less arbitrary than a form which is not. As mentioned earlier, the Aguaruna form for 'four' is particularly interesting not only because it is transparent in its meaning but
because it relates the names of numbers to the 'concrete'
action of reckoning.

The expression used for 'five' in Shuar is ewêh amus, 'the
hand is complete'. It is a descriptive sentence.

For numbers beyond 'five' we observe a good degree of variation in the expressions which are used together with the
gestures. Pellizarro (1969, p.23) gives a set of expressions
which are different from those given by Karsten and Ghinassi.
The numerals 'one', 'two', 'three', and 'four' are followed
by the verbal expression íraku, 'added'. Thus, one has
chikíchik íraku, 'one added', Jimerá íraku, 'two added',
etc., for 'six', 'seven', and so on.

The differences between the authors become still greater
when we go beyond 'ten'. For example, for 'eleven',

Pellizarro gives the expression chikíchik nawén íraku, 'one of the foot added'. However, he also states that "when they have more than four or five things to count they use to say untsurí 'many'". Both the divergence between Karsten's and Pellizarro's accounts, and the alternatives that Pellizarro gives, are important in the perspective I am presenting here. They confirm my field observations of the variation which is found after the first few established numerals. The variation is a central characteristic in the dynamic view of language and culture. These data bring insight into the growth of numerical terminology and demonstrate that the establishing of a numerical system can be a slow process in the history of a language. In this sense we can expect that some number names are connected with other roots of the language.

Recently a Shuar mathematical terminology was proposed as a part of a process of language standardization for a bilingual and bicultural education-by-radio program.[8] A group of Shuar teachers 'adapted' lexical items of the language to express elementary mathematical concepts. I already mentioned the root nekap-, used to express the meanings of 'number' and 'to count'. The form eweh, 'hand', was used for 'five'. Numerals for 'six', 'seven', and 'eight' were proposed. As for 'nine' the form usúmtai, which I have already mentioned, was introduced. For 'ten' the form náwe, 'foot', was proposed, to avoid expressions such as mái ewéh amúkhai, 'I completed both my hands'. For 'hundred' was proposed the form wašim and for 'thousand' the form nupantí, both being terms taken from the language. The entire numeral system is built on the base of these numerals and two basic syntactic principles of the language: the numeral which precedes the forms for 'ten', 'hundred', and 'thousand' has the function of a multiplier, the numeral which follows represents an added value. Thus, for example, himyára náwe čikíčik, 'two (times) ten (and) one', signifies 21 and menáintyu wášim áintiuk náwe ewéh, 'three (times) hundred (and) four (times) ten (and) five', signifies 345.

In addition to the numerals, approximately 45 terms were proposed for other concepts of elementary mathematics. I will mention some of these, to show both the linguistic problems and the internal possibilities that can be found in a native language. In the Shuar language there are various ways to express the disjunction. In mathematical terminology, the expressions niniák páčitsuk, 'not naming that', and tumátskeša or turútskeša, 'not doing this way', have been used. Such expressions are also found in the everyday language.

To express the concept of comparative order a current Shuar form was used. This is the plain adjective učič, 'small', for 'less than', and uúnt, 'big', for 'greater than'. To state, as many linguists would, that the Shuar language does not have comparative forms would mean that some standard model of comparative construction is taken as reference.

To express concepts such as 'less than or equal to', in which both the disjunction and the comparative are present, the disjunction was expressed in a way different from the translations already mentioned. In these cases it was simply omitted with učič', metéketai signifying 'is small, equal to', and uúnt, méteketai signifying 'is big, equal to'.

To conclude, I would like to emphasize once again the basic hypothesis of this paper. We cannot expect to find a specific mathematical activity in most cultures without a tradition of writing, but we should be able to identify basic capacities upon which a mathematical activity can be built.

In cultures without a written tradition, the whole perspective on knowledge and on the capacities of reflection is deeply different from our own. The process of accumulation of knowledge and of reflection on nature exists in a dimension very different from the one we know. Nevertheless, the language may "register" some aspect of such a process.

NOTES

1. I use _operative_ here in the Piagetian sense, focusing upon the abilities of planning without the need for trial-and-error manipulation, of executing the same action in both directions (with the consciousness that it is the _same_ action), and understanding the compensation of the dimensional change, characteristic of conservation.

2. This can be easily verified when we observe a discussion among mathematicians about a theorem or an equation. I mention this example not only because it is pertinent to our present object but also because recently the psycholinguist D. McNeill talked about the use of gestuality by mathematicians (Seminar on Cognitive Sciences, University of California, Berkeley, May 1981).

3. The Shuar of Eastern Ecuador number approximately 30,000 and are one of the largest indigenous groups of the Amazon. Together with the neighboring Achuar, Huambiza and Aguaruna groups they form one ethnolinguistic family (of which they are now aware) with more than 60,000 people in the Ecuadorian and Peruvian Upper Amazon. The name Jívaro is not accepted by the Shuar themselves but is still used in the literature; the ethnolinguistic family is referred to as Jivaroan. The linguistic field work for this study was carried out in 1968, 1970, 1971, and 1974.

4. In another, much more recent grammar, by another

missionary with a deep knowledge of the Shuar people we find: "the most frequent way is to count the values of ten, they do it showing the fists and saying atáksha 'again' until they reach the number they want." (Pellizarro 1969, p.24). Rivet (1907-1908) noted the little information available to authors of the XIX century: "Father Plaza [Compte 1885: II, p.195] says that they are able to count up to 4. Raimondi [1863, p.39] up to 5. Reiss [1880] writes that in Macas the savages use their fingers to show the numbers. The Aguarunas have numbers up to five and for the numbers beyond 5 they use the fingers and the toes [Hassel 1902, p.83]. The dialects of Macas, Gualaquiza and Zamora have words to express the first ten numbers; beyond 20 the Indians use the word irúnima that means 'numerous'. Everywhere they have been in contact with the colonists, they use the quichua numerals beyond ten."

5. See Note 3.

6. The Shuar of Ecuador are organized in a Federation whose main concern is the defense of land belonging to the native people. In order to achieve full economic independence and critical self-consciousness, the Federation operates a radio station which broadcasts in Shuar. For the education of the native children radio broadcasts have been used since 1972. In each jungle village a local teacher (referred to as teleauxiliar) follows the program as it is broadcast and will eventually adapt it to local needs. The Federation's education-by-radio program is defined as bilingual and bicultural. In the last few years four primers for mathematical education were written, geared for the first grade. The texts resulted from direct experience in teaching mathematics, the conceptual and linguistic parts being composed by the Shuar teacher Pedro Kunkumas (Nekapmarar' 1976-1978).

7. The "Upano tribes" are the northwestern Shuar. At present, in the Upano valley there is the Center of the Shuar Federation.

8. See Note 6.

4. The Calendrical and Numerical Systems of the Nootka

William J. Folan

INTRODUCTION

The Nootka are a Wakashan speaking group of fishing, hunt-in and gathering people that have lived in villages on the heavily wooded inner and outer coasts and up the inlets on the west coast of Vancouver Island, British Columbia for more than 4,000 years. Although the Nootka occupy a large area of the coast, the greater part of the information contained in this paper pertains to those from the Nootka Sound Region, here considered a central place among all Nootkan groups residing on the island. These were the people first con-tacted by Juan Perez in 1774 after he had been sent north in response to Spanish fears that Russian activity in what is now Alaska would jeopardize Spanish claims to the Pacific Coast of North America. The next notable visitor to the region was James Cook who arrived on March 29, 1778, almost two years out of England, on his third voyage of exploration for the British Admiralty. It is from Cook's and later voy-ages during the 18th Century, both Spanish and British, that we learn about the Nootkans' manner of conceptualizing time and how they made counts of such things as their ages, neigh-bors and the products of their region.

CALENDRICS

The Nootkans, like all people, were aware of the passage of time on both day-to-day and season-to-season bases, knowl-edge essential to their survival as fishermen, hunters and gatherers. Further, they were capable of estimating

quantities of such things as fish or people and, if necessary, of reducing these quantities to absolute numbers.

The need to divide a day, season, or a series of seasons
into discrete units was not equally compelling to all Native
American groups. Although the Nootkans were aware that definite periods of the day were more appropriate than others for
carrying out specific activities, as were different seasons
and phases of the moon, they did not define these periods,
seasons or phases with beginning or ending points based on
anything other than the occurrence or non-occurrence of a
single or a series of events in nature. Nor did the Nootkans
have a definite system for establishing a chronology for succeeding sets of periods, seasons or phases. They could not
talk about the Summer of 42 but could refer to The Winter of
Our Discontent.

Jose Marino Moziño observed that outsiders considered
Yuquot Nootkan chronology to be obscure, either because the
Europeans had difficulty understanding it or because the
Yuquot Nootkans were careless about arranging their "calendars". The Yuquot Nootkans had no concept of time comparable
to European centuries. By calculating the ages of two of the
oldest men in Yuquot (one was about 90 years old and the
other about 70), considering their memories of their grandfathers' tales and assuming 20 years between generations, it
was found that the old men were aware of wars that had
occurred slightly less than 200 years before 1792. Beyond
this period "all is lost in darkness" (Moziño 1913).

While Cook's ships were anchored in Nootkan territory for
only a few weeks, his men were unable to learn the Nootkan
language well enough to understand more than a few words and
phrases, much less their system of calendrics. Nevertheless,
they did record that the Nootkans divided time into moons

(Bayly 1776-1779).

Later on, Robert Haswell (1941, p.107) was apparently the first to list the various "moons" the Yuquot Nootkans recognized and to match them with the twelve European months. In his list, shown below, the meaning of the indigenous terms is taken from Philip Drucker (1951, pp.116-117), C. Knipe (1868, pp.69-70) and Moziño (1913).

Hiesekackomilth	January	Period of high winds and most snow
Weeyackomilth	February	Herring fishing moon
Hiakolmilth	March	Herring spawning moon
Enuckhechetermilth	April	Getting ready for whaling
Quahkermilth	May	Salmonberries moon
Tahahtakahmilth	June	Sting (berries) moon
Sahtsmilah	July	Wasp moon (?)
Eneecoresamilth	August	Dog salmon moon
Berrie Nahalth	September	Berries (moon)
Cheecakomilth	October	Time for splitting and drying salmon
Mahmee exso	November	Elder sibling (moon)
Cathlatick	December	Younger Sibling (moon)

According to Moziño, however, the Yuquot Nootkans divided the year into fourteen "months", each of twenty days, with a number of days frequently added to the end of each period. The number of added days varied from period to period and from year to year depending on the feature, usually a characteristic activity, that distinguished each period from the others. The chiefs decided when days should be added to any particular period and when the next period should begin, and

because these decisions were based on variable features,
uncertainty regarding Yuquot Nootkan calculations of time
always existed (Moziño 1913).

The first period of the Nootkan year, equivalent to July,
was Satz-tiz-mitl. Besides its twenty ordinary days, many
other days were usually added to it depending on the availability of the fish, such as halibut, tuna, cod, and bream
that were caught then. The following period, which partially
fell within August was Tza-quetl-chigl, but Moziño did not
describe it. Only a few days were added to it. Ynic-coat-
tzimil was the period for cutting down trees by setting fire
to the tree base. The month derived at least the first part
of its name from ynic, 'fire'. Fish were scarce during Eitz-
tzul, Ma-mec-tzu, and Car-la-tic which preceded winter.
Winter occurred during Aju-mitl, Vat-tzo, and Aya-ca-milks
(which probably should read Vaya-ca-milks) and ended near the
middle of February. Aya-ca-milks began near the middle of
February and was a time noted for sardine fishing. Birds,
including sea gulls, were abundant during Or-cu-migle and the
following period, Cay-yu-milks, was noted for great celebra-
tions of religious festivals as well as daily whaling expedi-
tions. An entire year's supply of whale grease would be
obtained then. Fruit, roots, shoots, leaves, and flowers
were collected daily during Ca-huet-mitl and Atzetz-tzimitl
ending more or less during the summer solstice (Moziño
1913).

In the following list attributed to Moziño, Nootkan periods were equated with the twelve European months. The meaning of the indigenous terms is taken from the same sources
used previously.

Vya-ca-milks   January   Herring fishing

Aya-ca-milks	February	Herring spawning moon
Ou-cu-migl	March	Geese moon
Ca-yu-milks	April	Religious festival and whaling
Ca-huetz-mitl	May	Salmonberry moon
Atzetz-tzimil	June	Fruit, roots, shoots, leaves, and flower collecting moon
Sta-tzimitl	July	Wasp moon (?)
Ynic-coat-tzimitl	August	Dog salmon moon
Euz tzutz	September	Rough sea moon
Ma-mec-tzu	October	Elder sibling (moon)
Cax-la-tic	November	Younger sibling (moon)
A-ju-mitl	December	Great cold moon (?)

Vat-tzo and Tza-quetl-chigl, which supposedly occurred during January and August, respectively, were omitted and the list is not as accurate as the one in Moziño's text which gives fourteen, not twelve, Nootkan time periods (Moziño 1913).

It was thought by others that the Yuquot Nootkans divided time into lunar months, ten of which formed one sun, and this division was considered to be derived from the human gestation period. In addition, it was said that divisions of a day depended upon the sun and the amount of fishing being done. Meal times and rest periods were also adjusted to the exigencies of fishing (Novo y Colson 1885; Anon n.d.a; Viana n.d.).

One Nootkan, Natzapi, had told the Spaniards his age at the time of Cook's earlier visit by repeating the Nootkan term for ten after each year. As a result, some Spaniards, like the writer of the Canto de Alegria (Anon n.d.b), may

have incorrectly inferred that the Nootkans used a solar year
with ten divisions and employed a decimal system (see also
Anon n.d.a; Viana n.d.). The writer of the Canto de Alegria
attempted to describe how the Nootkans told him they determined time but the result is almost incomprehensible:

"El Arrumban.^{to} de su Plano deve ser proximam.^{te} el N.S.
Nos lo esplico manifestando, que el Sol salia si en diferent.^{s}
parajes (segun su Declinac.^{n}) pero que siempre en unos Circu-
los perpendiculares a esta linea corria de la derecha a la
isquierda; pudiera inclinarese algo a el O, porque dia, que
el Sol antes nace es Nutka, y Tasis, que en los Muchimases:
Gradua por un dia de Camino la dist.^{a} desde Nutka a Tasis."

Thanks to Jack Himelblau, University of Texas at San
Antonio, we are able to present the following translation:

"The bearing of ... [the sun's] plane seems to be approximately north-south. He explained it to us stating that the
sun rose in different places (according to its declination)
but that it always moved in circles perpendicular to this
line from the right to the left; it could incline somewhat
toward the east [or west] because the day in which the sun
first rises in Nootka and Tahsis [to the east of Nootka or
Yuquot] rather than in Muchimases [on the northeast coast of
Vancouver Island], it graduates by one day on the way from
Nootka to Tahsis."

During the late historic period, the Nootkan measurement
of time was a yearly cycle divided into a two phase unit calculated by placing a stick in the ground to observe the type
of shadow it cast to determine whether the sun was measuring
in a northerly or southerly declination. Lesser periods were
computed according to lunar periods, twelve or thirteen to
the year. The lunar count primarily referred to natural

phenomena such as fish runs or flights of water fowl, as
early historic writers observed. Drucker (1951, pp.115-116;
n.d.) found that moon counts of the Northern Nootkan groups,
such as that published below, were almost identical. The
association with European months is only approximate.

Wīyàqhàmɬ	January	No (food getting) for a long time (?) moon
Axhàmɬ	February	Bad weather moon (?)
Ai'tamiɬ	February	False (spawning) moon
Aiyakàmiɬ	March	Herring spawn moon
Hō'ukamɬ	April	Geese moon
TaqLatqɔàmɬ	May	Stringing (berries) moon
Qawɔcàmiɬ	June	Salmonberries moon
Asatsàmiɬ	July	Wasp moon
Satsàmiɬ	August	Spring salmon (run) moon
Hɛniqoɔàsàmiɬ	September	Dog salmon moon
Etsosimiɬ	October	Rough sea moon
Ma'mīqsū	November	Elder sibling (moon)
Qaɬatik	December	Younger sibling (moon)

When comparing Haswell's and Moziño's moon counts, presumably collected from Nootkans living at Yuquot, a moon count
published by Knipe and also found in Sproat (1868), a moon
count Drucker collected from an Ahousat respondent, and
Drucker's published version, much of which was collected
from a resident of Yuquot, one can readily understand why
Drucker found accounting for the differences between them
difficult. Drucker thought one would expect all moon counts
to differ or all to be uniform instead of being uniform among
the Northern Nootkans and varied among their southern neigh-
bors. Actually, if any faith can be placed in the

ethnohistoric record, the terms of the various lunar months
may have changed through time, at least in the Nootka Sound
area.

For example, the term Haswell associated with January dif-
fered from the corresponding term Moziño and Drucker collected, but agreed with the term Knipe published. Haswell's
term for February agreed with Moziño's, but disagreed with
those of all other writers recorded. Haswell's term for
March was the same as that noted by Moziño, Drucker's Northern Nootkan respondents, and Knipe, but Haswell's term for
April did not agree with anyone else's. Drucker's term for
April fairly closely matched Knipe's, and so on.

Some of the differences are readily explicable. For example, the northern Nootkan term corresponding to January, as
collected by Drucker, indicated that no food could be
obtained for long periods at this time, a well-documented
historic fact; Haswell's and Knipe's term for January roughly
meant that high winds and the most snow occurred then. Such
weather conditions would make obtaining food very difficult
and the apparent anomaly was merely a difference of emphasis:
one emphasizing cause and the other, effect.

Haswell recorded the term Enuckhechetermilth for April and
one of Drucker's Clayoquot respondents gave Inihiekmił for
May. Both referred to preparing for whaling. Drucker's
Northern Nootkan respondents did not refer to whaling at all
in any of the terms they gave for the thirteen lunar months;
April was "Geese moon" and May was "Stringing (berries)
moon". This too is explicable. Between the time Haswell was
on the coast and the time Drucker was there, whaling had lost
much of its importance and the Northern Nootkans may have
switched from emphasizing whaling to emphasizing berries.

The lunar counts agreed as to the relative position and

time of year that the elder and younger sibling moons
occurred. The one exception was the moon count an Ahousat
respondent gave Drucker.

DISCUSSION

In general, all the moon counts make sense. The major
differences were due to early recorders misquoting the chronological order of the lunar months or trying to fit the thirteen or fourteen Nootkan "months" into our twelve month
system. Also contributing to present-day confusion regarding
Nootkan calendrics were the different emphases placed on the
various characteristics of each period by different Nootkan
groups through time, including a few present-day residents of
Yuquot.

Today, most Nootkans own watches, hang calendars on their
walls, and talk about such definitely dated events as the
hard times of the '30's. They are aware of what an eight-
hour day is and that overtime is paid at one-and-one-half
times their normal wages. Payday is, of course, well marked
and if it passes without the customary work-reward, the time
the pay is overdue is readily calculated. Boat schedules and
other such timed events are also adhered to. Nevertheless,
both the people living at Yuquot and their non-Indian neigh-
bors still distinguish between "Indian time" and "white man's
time" when establishing a particular hour to do something,
indicating that the difference between them is not only real,
but remembered.

NUMERICS

Almost all word lists formed in the Nootka Sound area contain Nootkan terms from one to ten or higher for rather obvious reasons. Most recorders would be interested in obtaining

fairly exact answers to the questions they prefix by phrases
such as "How many ..." or "How much ..." rather than receiving general replies like "A lot", or "Not very many", or
other such indeterminate quantities. Besides this, it seems
easy to record something as apparently basic as numbers.
However, this was not always the case with the Nootkan form
of counting and many early journalists went astray in recording Nootkan numbers.

The Yuquot Nootkans were thought during the contact period
to do much of their reckoning in tens indicated by clapping
or clasping their hands together, thereby eliminating the
need to express ten verbally. When they counted higher than
ten, they always used the same terms to express units from
one through nine, then clapped or clasped their hands. When
they clapped twice, James Strange understood them to mean
twenty, three claps meant thirty, and so on (Strange 1928,
p.54).

Strange recorded no calendrical terms, but did record
numerical terms (1928, pp.53-54), as did Anderson and Burney
before him (Burney n.d.; Cook 1967, p.330). According to
Strange these were as follows.

1	Sauwāāk
2	Ahtkla; Akkla
3	Kutsa; Katsāā
4	Moo; Mo; Moatla
5	Soocha; Socha
6	Noopo; Noopokh
7	Alkhpoo; Atlpo
8	Atlaqualkh
9	Souwaqwalkh
10	Haēēēo

11	Saoometeepahaēēēo
12	Ahlklemehapahaēēēo
13	Kutsamelepahaēēēo
14	Moomahtehapahaēēēo
15	Soochamehtepahaēēēo
16	Noopomehtehapahaēēo
17	Athpomehtehapahaēēēo
18	Atlaqualkhmehtepahaēēēo
19	Sowaqualkhmetepahaēēēo
20	Sakaits haēēēo
30	Haeeemehlepatsa keets haēēēo
40	Haēēēo Akkleook
50	Haēēēo metta putta akkleook; Kutseatlish haēēēo
60	Haēē mehlaputkutseak haēēēo
70	Mooeeak Haēēēo
80	Soocheak haēēēo
90	Haēēē metla put soockeak haēēēo
100	Noopock
110	Haēēēmehtla put noopok haēēēo
120	Atlpok haēēēo
130	Haēēēmehtla put soocheak haēēēo
140	Atlaqualkhuck haēēēo
150	Haēēēmehtla put atlaqualk huk haēēēo
160	Sowaqualkhuk hāēēeo
170	Haēēēookh
180	Sukkytzuk haēēēo
190	Haēēēmehalputs sukkytz haēēēo
200	Atlepok haēēēo

Although the above list strongly suggests a vigesimal system, neither Strange nor others seemed to recognize this,
including Moziño (1913) who stated that the Yuquot Nootkans

had a particular word for all numbers from one through ten; twenty was expressed as twice ten; thirty, as three times ten; and so on. The Yuquot Nootkans were not thought to use exact figures when referring to quantities in the thousands, but represented such quantities indefinitely by repeating the word for ten from five to seven times. Malaspina incorrectly thought the Yuquot Nootkans were unable to count above ten and had difficulty expressing greater numbers without the aid of some physical device such as a finger. For example, the Yuquot Nootkans used this technique to tell the Spanish their ages and to indicate time in terms of "suns" (Anon n.d.b; see also Moffat n.d.).

Augustin J. Brabant (n.d.) and Edward S. Curtis (1916) presented a much more complex picture of the system of counting for the Northern Nootkans than described in the early ethnohistoric record (see also Jewitt 1807; Sproat 1868). Brabant's list of number words for the Hesquiat Nootka is published below, together with an analysis of the structure of the compound terms.

1	Tsawoik	
2	Atla	
3	Katstsa	
4	Mo	
5	Socha	
6	Noupo	1 (on the second hand)
7	Atlpo	2 (on the second hand)
8	Atlakwoitl; Atlakwol	2 lacking (to 10)
9	Tsawoikwoitl	1 lacking (to 10)
10	Haio; Hayo	
15	Hayo ogish socha	10 and 5
	Socha matlap hayo	5 over 10

18	Hayo ogish atlakwol	10 and 8
	Atlakwol matlap hayo	8 over 10
19	Hayo ogish tsawoikwol	10 and 9
	Tsawoikwol matlap hayo	9 over 10
20	Tsakets	1 score
30	Tsakets ohish hayo	1 score and 10
40	Atleek	2 score
50	Atleek ohish haio	2 score and 10
	Hayo matlap atlek	10 over 2 score
60	Katseek	3 score
70	Katseek ohish hayo	3 score and 10
	Hayo matlap kachtsek	10 over 3 score
80	Moyek	4 score
100	Sochek	5 score
120	Noupok	6 score
140	Atlpok	7 score
160	Atlakwoitlek	8 score
180	Tsawoikwoitlek	9 score
200	Hayok	10 score
300	Hayok ohish sochek	200 and 100
400	Atlpitok	2 × 200
800	Mopitok; Mopit hayok	4 × 200
4000	Tsaketspitok	20 × 200

DISCUSSION

Although most of Strange's information on the Nootka is fairly accurate, he apparently made some errors while recording numerical terms. His entries are generally cognate with those of the Northern Nootka but exhibit some variance with the later records of Brabant (n.d.) and Curtis (1916). For example, Strange's rendering of 11 and 12 is in disagreement with that of Curtis, but their formation is consistent with

the second formation found in Brabant's listings for 15, 18,
and 19. Strange's terms from 20 on appear to contain super-
fluous words for 10 (haēēēo) and, in addition, are mismatched
with Brabant's terms from 60 on. It is difficult to determine why Strange broke cadence between the English and Nootkan systems of numeration but my guess is that as soon as he
ran out of fingers it became more difficult for both him and
his Nootkan respondent to comprehend what quantities they
were trying to record.

Brabant firmly contradicted early inferences for the
Nootka Sound area that the Nootkans possessed a decimal system and stated that they used a vigesimal system. He also
confimed that they used their fingers to count and gave the
Nootkan system of using different terms for different
objects. For example, the Nootkans use different terms for
counting or speaking of:

a) people, men, women, children, salmon, tobacco

b) anything round in shape such as the moon, clothing
(except for trousers), birds, vessels, etc.

c) anything long and thin or narrow, such as rope or
trousers

d) an object containing many things such as a block of
matches, a herd of cattle, a bale of blankets, etc., and
several other classes of things.

The Nootkans used various aids to help them to remember
numbers. They customarily tied knots in a string to keep a
record of the passage of lunar months. In this way could be
rcorded the number of times a man had performed a particular
bathing ritual, how many sea otters a hunter had killed, how
many days a trip had taken, or the number of days a pubescent
girl had spent in restriction. At potlatches, bundles of
sticks were used as memonic devices indicating how many

chiefs and lower ranking men had been invited (Drucker 1951, pp.116-117). One thing made clear in both the early and the more recent ethnographic record is that the numbers four and ten are often used in close association with supernatural events, but why this is so is not known. For example, during a whaling ritual a chief had to ceremonially bathe in a lake by rubbing himself with one type of plant or another for four nights and to later walk around the lake during the next four nights in preparation for the whaling season. If a whale is harpooned and beached, however, the chief's skill may be attested to ten spirits (ya'ai) often associated with whaling related activities (Drucker 1951, pp.171, 179).

ACKNOWLEDGEMENTS

This paper represents a revised version of a chapter from my unpublished manuscript on Yuquot, Where the Wind Blows from All Directions: The Ethnohistory of the Nootka Sound Region and is published here through the courtesy of Mr. John Rick, Chief of Research, Parks Canada, Department of Indian Affairs, Ottawa. I would also like to thank Jean Brathwaite and Michael P. Closs who read and improved the manuscript. Any errors or omissions are, however, exclusively those of the author.

This paper was written while I was an adjunct faculty member at St. Patrick's College, Carleton University, Ottawa, Canada. I would like to thank Dr. Gordon Irving for the considerable courtesies extended to me and to members of my staff while at the College. Likewise, I wish to acknowledge the financial support provided by Canada Council Grants 68-1550, 568-1550-51, 570-0557 and 570-0557-51, as well as monies granted me by the National Historic Parks Service, Department of Indian Affairs and Northern Development, Canada.

5. Chumash Numerals

Madison S. Beeler

Chumash is the label identifying a family of languages
spoken in aboriginal times along the coast of southern California from about Malibu, northwest of Los Angeles, to an
indeterminate point north of the city of San Luis Obispo.
They also occupied the three westernmost of the Santa Barbara
Channel Islands: Santa Cruz, Santa Rosa, and San Miguel.
They were a coastal people; but we know that in the general
region of Ventura they lived beyond the summit of the mountains bordering the southern end of the San Joaquin valley.
So far as is known there are no speakers of any Chumash language alive today; but in the nineteenth century Indians gave
us some information about seven different forms of Chumash
speech. It is likely that before white contact there were
still others.

   In the late eighteenth and early nineteenth centuries five
missions of the Franciscan order were founded among these
people. A generalized form of language tended to grow up at
each mission center, and these dialects of languages are traditionally known in the literature by adjectives derived from
the Spanish names of the missions. There is thus, commencing
in the south, a Ventureño dialect; to the northwest this is
succeeded by the territory of the Barbareño -- city of Santa
Barbara -- group. Next are Ynezeño -- mission of Santa Inez
-- and Purisimeño -- mission of La Purisima Concepcion [the
Immaculate Conception]. The last group to the north are the
Obispeño, from the city and mission named for San Luis
Obispo. No mission was established on the islands, or in the

interior; the forms of Chumash speech there are usually known as Cruzeño, from Santa Cruz, the name of the largest island, and Interior.

My interest in the study of Chumash speech began almost thirty years ago when a fluent speaker of Barbareño was discovered in Santa Barbara. At that time almost nothing about these languages, beyond a few poorly recorded vocabularies, was available in the literature. This woman, Mrs. Mary Yee, had been born in 1897 and was then in her middle fifties. I worked with her, on and off, until her death in 1965. The result of this work, and of that of a number of students, is that we now have grammars and dictionaries, not only of Barbareño, but of most of the other Chumash languages. This rescue operation was carried out at the last possible moment, since there now appear to be no speakers of any of these languages. This material now makes possible linguistic work of many kinds on these fascinating languages.

In 1961, while working in Santa Barbara, I had called to my attention by Father Maynard Geiger, O.F.M., the resident historian of the Franciscan order, a small manuscript book in the order's archives. This turned out to be a 'Confesionario', or handbook for father confessors, in Ventureño, with interlinear translation into Spanish, and some passages in Latin. The handwriting of this document was identified for me by Father Geiger as that of José Señán (1760-1823), a native of Barcelona who was stationed at Mission San Buenaventura from 1797 until the end of his life. Contemporary testimony of his ecclesiastical superiors emphasizes Father Señán's "knowledge of the languages of it's (i.e. of the San Buenaventura mission's) Indians." This document I published in 1967 (Beeler 1967). Earlier, however, separate publication was given to a unique section of this handbook

which had attracted my attention because of my interest in
the study of aboriginal counting systems (Beeler 1964). This
part of Father Señán's work had great importance for me
because of its early date. I knew something before about
Chumash numerals; but all of what I had learned before was
outdated by the information contained in Señán's handbook.

Most of the facts previously available about Chumash
counting had been recorded in the second half of the nineteenth century or early in the twentieth. It appears that
one of the parts of native grammar most vulnerable to intrusive influence was precisely the system of numerals. The
padres did not wish to adjust their thinking to accommodate
anything differing from the decimal system to which they were
accustomed by their own Indo-European language. Even linguistically more sophisticated investigators have difficulty
in divesting themselves of ingrained habits, and many of them
wanted to impose these habits on the Chumash when they began
to become familiar with the strange counting practices of
these Indians.

A passage in Señán's handbook illustrates these points.

"Quest. To how many have you said that what the
          Father says is a lie?

Ans.   To fourteen (a catorze, eshcóm laliét).
          This expression in Ventureño means something
          like 'two lacking, subtract two'.

Quest.   I don't understand what you say to me (no
          entiendo lo que me dices). I don't under-
          stand the way you people count (no entiendo
          vuestras cuentas): count by tens (cuenta por
          diezes, saliét al cashcom).

Ans.   Ten and four (diez y cuatro, cashcóm
          casatscumu)."

It may be guessed that this passage was written about 1818-1820, which is very early as Indian language texts in California go. The reader will soon see that the Ventureño manner of counting was apparently flourishing in its aboriginal condition at that date. Subsequently, in this paper, he will see that influence of this kind, wielded by the priest in the confessional, had effectively destroyed the native system of counting at an early date: by the second half of the century the Indians were all "counting by tens." Most California native languages apparently became known to linguists only after influence of this kind had done its work: the native system had been lost irretrievably before the investigator ever saw it. It is this fact that gives the Señan record its great value for science. The Ventura mission had been founded in 1782; the native system of numbers had survived for some thirty five years, but would soon succumb to the treatment it was receiving.

It is time now to examine this native system. This can most easily be accomplished by presenting Señan's description, with some comments by me. The spelling is that of the original.

1	paqueet
2	eshcóm
3	maseg
4	scumú
5	itipaqués
6	yetishcóm
7	itimaség
8	malahua
9	etspá
10	cashcóm

11   telú
12   maseg scumu, tres vezes cuatro
          [three times four]
13   masegscumu canpaqueet, tres vezes cuatro [sic;
                              three times four and one]
14   eshcom laliét, dos faltan pạ 16
          [two lacking from 16]
15   paqueet cihue, uno falta pạ 16
          [one lacking from 16]
16   chigipsh
17   chigipsh canpaqueet
18   eshcóm cihue scumuhúy
19   paqueet cihue scumuhúy
20   scumuhúy
21   scumúhuy canpaqueet
22   eshcóm cihué, dos faltan pạ etsmajmaség
23   paqueet cihué, uno falta pạ etsmajmaség
24   etsmajmaseg
25   etsmajmaseg canpaqueet, veinte y cuatro y uno
26   eshcóm cihué itimaseg, dos faltan pạ veinte y
                              ocho
27   paqueet cihue itimaseg, uno falta pạ veinte y
                              ocho
28   ytimaseg,] maseg
29   ytimaseg canpaqueet, veinte y ocho y uno
30   eshcom cihue eshcom chigipsh, dos faltan pạ dos
                              vezes diez y seis
31   paqueet cihue eshcóm chigipsh, uno falta dos
                              vezes diez y seis
32   eshcóm chigipsh, dos vezes diez y seis

Desde el numero 32 empiezan a contar de cuatro en cuatro,

como sigue:

scumú.................... 4
malahua.................. 8
maség scumu.............. 12
chigípsh................. 16
scumuhúy................. 20
etsmajmaseg.............. 24
ytimaseg................. 28

[Here there is a break in the manuscript; when it begins
again, on the next page, we have:]
el numero 16, que sale siempre duplicado al fin de la
cuenta. Asi aora prosiguiendo en contar, el numero ultimo
sería yetishcóm chigípsh; esto es, seis vezes 16. Pasando
adelante, sería malahua chigípsh ocho vezes 16; y asi de los
demás. [the number 16, which always comes out duplicated at
the end of the count. Now, continuing the count, the last
number would be yetishcóm chigípsh; that is six times 16.
Still going ahead, malahua chigípsh would be eight times 16;
and the same for the rest.]

In the discussions that follow I shall assume that the
reader is familiar with the names of the Chumash languages,
other than Ventureño, which have been listed above. These
languages are related to each other as follows: Ventureño is
a member of a cohesive group of four, quite similar to each
other, called Central Chumash, and containing, besides
Ventureño, Barbareño, Ynezeño, and Purisimeño. The dialect
of the islands, which we call from its principal variety
Cruzeño differs in many respects from the Central Chumash
languages; and the language of San Luis Obispo to the north,
called Obispeño, is the most divergent of all six. In

aboriginal days there assuredly existed other forms of Chumash speech; but missions were not founded among them, and
very little is known of these Indians. I shall use the term
'Interior' when referring to Chumash groups who lived away
from the Pacific coast.

A recurrent feature characteristic of the history of
native numerals in California is the prevalence of borrowing
from language to language and from dialect to dialect. What
is borrowed is not only the number words themselves (the
numerals), but also, occasionally, the basic system itself.
Before the coming of the whites, the California region was an
area of great linguistic diversity. Because of this, and
because many forms of native speech have become extinct without ever having been recorded, it is not always possible to
identify the source when borrowing is suspected. But it frequently is, and I give this information when it is known. It
will be useful to have available a map of California, if you
wish to understand the geography of what is involved.

The arithmetical operations employed by the Ventureño, to
construct their system, will be seen to be addition, subtraction, and multiplication. All numeral systems with which
this writer is familiar employ a restricted number of root
words; these root words are then combined with each other,
and modified with affixation, to express all the numbers
which the culture in question needs. The analysis of such
systems requires then, the isolation of the root numeral
words, the description of the processes by which these root
words are brought together to express higher numbers, and the
identification of foreign borrowings, if any appear to have
been made.

The root words of the Ventureño are basically four in
number: they are the terms for 1, 2, 3, and 4, i.e. (in

Ventureño) paqueet, eshkóm, maseg, and scumu. Hereafter, in
this paper I shall replace these spellings, which conform to
Spanish orthographic conventions by the following: pake?et,
?iškóm, masix, and skumu, which follow current linguistic
usage. Any interested reader can determine the meaning of
these symbols by consulting an elementary textbook of phonet-
ics. The system is obviously quaternary; the only other root
words employed are the words for multiples of four: malawa
is eight, and pet'a is sixteen (in Barbareño). We shall soon
discuss the Ventureño term for this number, šixipš, a Ventureño innovation.

The expressions meaning 5, 6, and 7 will be seen to contain the words for 1, 2, and 3, prefixed by an element yiti-,
which from the exigencies of the system seems to mean 'four';
five would be 'four plus one', six 'four plus two', and seven
'four plus three.' I was once inclined to connect this element yiti- with a Yokuts verbal root yit'is, 'to make five'
(Beeler 1976, p.256). (Yokuts is a major Californian family
of languages spoken, beyond the coastal mountains from Chumash, in the San Joaquin valley. We shall encounter potential Yokuts influence repeatedly; that transmontane tongue
appears as the source for the introduction of quinary and
decimal features into Chumash.) Subsequent considerations
have led me to prefer a native Chumash etymology for yiti-.
We find in some members of the family a verbal root yet'i ~
yit'i-signifying 'to come, come back, return.' We have in
Cruzeño (see below), for 5, 6, and 7, constructions such as
(na-)syet-eshkom, '6', which I think can take an interpretation 'when (or where) two recurs, is seen again'. In principle I prefer a native etymology to a foreign loan, if I can
find one.

Since the number 5 is expressed as the sum 'four plus

one', we might expect, in a quaternary system, that nine
would be given as 'eight plus one.' The Ventureño for 'nine'
(in Señán etspa, in my notation tspa) may be given such an
interpretation, in the light of the whole Chumash system.
The central Chumash for 'one' commences with the syllable
pa-; if this be interpreted as a verb 'to be one', as are so
many words in Indian languages, the Ventureño third person
singular subject pronoun, when prefixed, yields tspa, 'it is
one.' This of course requires the assumption that prehistoric phonetic changes have, under unknown conditions, caused
the reduction of pak'a -- or whatever it once was, to pa.

Whatever the unknown -- and unknowable -- prehistory is of
this word for 'nine', it is certain that Barbereño used an
expression for 'ten' meaning 'add two.' This is clearly its
term for 'ten', k'eleškóm, which shows -eškóm, 'two', following the segment k'el-, which may be translated 'and'. 'Ten'
is therefore rendered as the sum '(eight) and two'. In
Ventureño kaškom we find the same construction, somewhat
obscured by subsequent phonetic change. The Ventureño for
'and' is kal- ~ kan-, which, when prefixed to eskóm, caused
elision of the initial vowel of 'two' and later reduction of
the resulting consonant cluster.

What I can suggest for til'u, 'eleven', is much less certain. One suspects here something signifying 'three', but
one must reach far to find a possible source. Nothing
remotely similar has been discovered in the areas which we
know have supplied linguistic loans to Chumash. The only
sources possible, when phonetic and semantic structure are
taken into account, are in the San Francisco Bay region, more
than five hundred km away to the north, and in Polynesian,
where we find a word something like telu, in the sense of
'three'. I will urge the case of neither. Only one Chumash

language, Obispeño, has a word for 'eleven' not cognate with
the Ventureño.

The expressions for 13, 14, and 15, according to Señán,
are, respectively, '12 plus 1', 16 less 2', and '16 less 1'.
These show the importance, in this system of counting by
fours, of 12 and 16, multiples of 'four'.

The next critical point is 16, the product of the basic
'4' when multiplied by itself. There is in several Chumash
languages a root word for this unit, but that does not occur
in Ventureño; there we have chigipsh (in linguistic spelling
šixipš). For it a convincing etymology has been discovered.
In the Central Chumash languages there is a verbal root iquip
~ ixip, meaning 'to shut, to conclude, to complete'; this is
here preceded by the third person singular pronominal prefix
s- and followed by the intensive suffix -š, and the whole is
realized by the characteristic Chumash process of sibilant
harmony to yield the existing form. It means then '(it, the
count) is quite complete, is concluded'. It appears to be a
Ventureño innovation, possibly to supply a likely meaning to
an inherited term no longer understood. Later in this paper
we shall see what subsequent generations did with it.

No comment is called for about the numerals for 17-19,
21-23, 25-27, or 29-31. For 20 we have scumuhuy; this shows
scumu, '4', provided with a suffix -huy, of unknown meaning.
Because of the words for 24 and 28, to be examined directly,
I suspect a significance of 5 for scumuhuy, that is,
'5(x 4)' = '20'; -huy would then be a term for 'one'. For
28 there is yiti-masix, which is simply the word for 7.
Therefore in tsmax-masix for 24 we should have an expression
for 6; since masix is 'three', tsmax- could be a term for
'twice', otherwise unknown. Or, one might admit other
possibilities.

The system is, then, a consistently elaborated pattern of counting by fours as far as 4 × 4; when 16 was reached, the process of counting as far as 4 × 4 was repeated. From 32 on the whole doubled quaternary count was repeated, with no upper limit stated.

As remarked earlier in this paper, the Ventureño count is the only one of the Chumash family to be worked out so thoroughly, and that is because of its early recording and because of the pains taken by Father Señán to make this recording complete. Most of the other Chumash languages begin to show the interference of the Spanish decimal system after twelve, as well as the employment of loan words from Spanish. Some of these innovations will be noted below.

We now pass to the consideration of counting in the other Chumash languages, and we look first at Barbareño, the coastal tongue just to the west of Ventureño, and a member, as stated, of the central Chumash group. This language continued to be spoken until 1965, much longer than any other idiom of the family. I can therefore give the numerals as I heard them spoken in the twentieth century.

1	pak'a
2	ʔiškóm'
3	masix
4	skum'u
5	yitipak'a
6	yitiškóm'
7	yitimasix
8	malawa
9	spaʔ
10	k'el-eškóm'
11	t'ilu

    12   masixeskumu
    16   (s)pet'a

All other numerals, when required, were borrowed from
Spanish. The term for 16 appears to be the native Chumash
word, replaced in Ventureño by šigipš. pak'a is a slight
variant of Ventureño pake?et. All the rest have been discussed in the Ventureño section.

There are two other central Chumash tongues, Ynezeño and
Purisemeño; named after the missions at Santa Ynez and La
Purisma, less than 25 km apart. These can be quickly
disposed of. The data are given below.

	Ynezeño	Purisemeño
1	pakas'	kac' (i.e. kats')
2	?iškom'	?iškom'
3	masix	masax
4	skumu	tskumu
5	yitipakas'	tip'ak'ac'
6	yitiškom'	te?škom'
7	yitimasix	tem'asax
8	malawa	malawa
9	spa	cpa
10	č'iyaw	kac'ač'iyaw
11	tɨl'u?	(no others given)
12	xayiskumu	
13	k'elpakas'	
16	pet'a?	

First the Ynezeño forms. Here 'one' has a slightly variant shape, when contrasted with Ventureño and Barbareño. The

only other numerals requiring comment are č'iyaw, '10', and
xayiskumu, '12'. The first of these is surely a borrowing
from the Yokuts of the southern end of the San Joaquin valley, where the Yokuts dialect called Yawelmani has t'iy'e•w
for 10 (Newman 1944, p.55); Yokuts numerals were decimal.
There is a history of intermittent contact between these
southern San Joaquin Yokuts and the coastal Chumash, and when
the Chumash of some of the missions revolted against Spanish
(or Mexican) rule in 1824 it is to this region that many of
them fled. One also assumes trade between these groups. As
for xayiskumu, '12', it is said that xayi- here means 'and'
(Applegate n.d., p.185); 12 was '(8) and 4'; this is the only
central Chumash term for 12 thus far encountered which does
not have '3 × 4' for 12.

Purisimeño is one of the most imperfectly known of all
Chumash dialects; what I here give, the sequence of the first
ten numerals, is supplied through the courtesy of Kathryn
Klar from records at present in the National Anthropological
Archives in Washington, D.C., and collected by John P.
Harrington. These show a system basically identical with
Ynezeño, differing from it only in phonetic details. Of the
latter, the most characteristic is the loss, under unknown
conditions, of the initial syllable in some words. The word
for 10 is a compound signifying 'one ten', with the Yokuts
loan as the basic term.

I come next to Cruzeño, or Island Chumash. Those three of
the Channel Islands occupied by Chumash speakers are reported
by 16th-18th century explorers to have been relatively
densely populated. There were, however, no Indians at all
left on them by the middle of the 19th century; it appears
that the Island people were either induced by missionaries at
the mainland establishments to abandon their island homes, or

were subjected to harassment by Aleut sea-otter hunters
imported by the Russians. No missions were ever founded on
any of the islands. What knowledge we have of island speech
comes from data collected in the latter part of the 19th
century from surviving speakers on the mainland. The most
copious of these records is the last one, collected in 1913
by John P. Harrington from a speaker said to have been born
on Santa Cruz and taken to Ventura ca. 1808; he is reported
to have been 109(!) years of age when Harrington worked with
him. He had lived most of his life in Ventura; Ventureño is
said to have been his customary speech. What follow are the
numerals which this man gave to Harrington.

1	ismala
2	iščom
3	masɨx
4	skumu
5	(na)syet'isma
6	(na)syet'iščom
7	(na)syetmasɨx
8	malawa
9	spaʔa; tspa
10	kaškom
11	tɨl'u
12	masɨx(pa)skumu
13	masɨxpaskumu hɨwan ismala
20	iščompaška(ʔa)škom
100	kaʔaškòmpaš kaʔaškòm

Cruzeño is the only Chumash dialect which does not show
for 'one' some form of the stem pak-. I am inclined to think

that isma(-la) may best be thought of as an original Chumash
expression for that numeral, replaced in all the other
related dialects by an importation from some unknown point on
the coast of southern California. This hypothetical importation lacked the impetus to carry itself across the twenty
miles of open water between the islands and Hueneme, their
port on the mainland. In support of my contention I cite two
neighboring forms of coastal speech, one to the south (Gabrielino, the native language of the Los Angeles area, of Uto-
Aztecan affiliation) and one to the north (Esselen, spoken
just south of Monterey, of probable Hokan character). The
word for 'one' in Gabrielino was pukú, in Esselen pek. The
attentive reader will have noted that the central Chumash
forms of 'one' exhibit varying patterns of suffixation to a
stem pak-, as if speakers found difficulty in assimilating a
strange loan word. Gabrielino and Fernandeño are the only
members of the vast Uto-Aztecan family which show a word for
'one' like pukú.

For 'five', 'six', and 'seven', variants recorded by Gould
in the 1880's exhibit a prefixed particle na lacking in Harrington's text. As in central Chumash these words contain
the terms for 1, 2, and 3; but the preceding element syet',
corresponding to the central Chumash yit'is, here has the
prefix s-. This looks like the 3rd person singular marker of
the subject. The particle na- in this language is a subordi-
nator signifying 'when'. The whole can then be interpreted
as meaning 'when, where 'one' recurs', or something similiar.
'Twelve' is of course 'three fours', and 'thirteen' means
'12 + 1', maintaining the old quaternary system. But 20 is
'2 tens' and 100 is '10 tens' showing the European influence
which is expected in the twentieth century. Perhaps, also
kaškom for 10 has undergone Ventureño shaping.

As stated above, Obispeño or Northern Chumash is the most divergent form of Chumash speech. The numerals there are the object of a recent study (Klar 1980; see below). The numerals are here quoted from the publication of Thomas Coulter, an English scientist who was in California in 1832-34; the record is slightly edited, to eliminate obvious typographical and other errors. The recording is so early (but was made about sixty years after the founding of the mission at San Luis Obispo) that no European influencing can be detected. Coulter was so perceptive that he ended his record with 'sixteen' in contrast with many later (and earlier) workers who wanted to find a decimal system in Chumash.

1	tskhumu
2	eshiu
3	misha
4	paksi
5	tiyewi
6	ksuasyu
7	ksuamishe
8	shkomo
9	shumochimakhe
10	tuyimili
11	tiwapa
12	takotia
13	wakshumu
14	wakleshiu
15	waklmishe
16	peusi

These words, strange as they at first seem, soon reveal the underlying quaternary base. In 6 and 7 we can discover

compounds containing 2 and 3; in 13, 14, and 15 we have further compounds with 1, 2, and 3. 16 is a root word, possibly
related to the pet'a of the central dialects. 9 probably
commences with 1.

In Klar's discussion, the Obispeño for 1, not elsewhere
present in Chumash, is attributed to a borrowing from Uto-
Aztecan, where a similar term is the standard word for this
numeral. We have seen the ubiquitous phenomenon of borrowing
in Chumash numerical sequences (and will see further examples), and must concede the possibility. According to Klar,
when tskhumu displaced paksi from the 'one' slot, this paksi
was transferred to the next critical point, the 4 slot, and
the term for 4 was moved ahead by four steps to 8, here dis-
placing the otherwise universal (for Chumash) malawa. This
is all ingenious, and so far as we can tell may well be
true.

In the numerals for 5 and 10 we see apparently non-Chumash
words, that is, borrowing. For Klar, tiyewi derives from
Buena Vista Yokuts, where it means 10 and was claimed as the
source for Ynezeño č'iyaw, '10'. If this theory is not
accepted, for semantic and/or phonetic reasons, we could fall
back on the assumption of unknown origin; in any event, both
this word and that for '10' illustrate the invasion of the
Chumash system by a quinary or decimal based sequence. As
for tuyimili, 10, no source has been found, unless we recognize such in the Esselen tomoila, '10' (the Obispeño word has
usually been taken to be the source of the Esselen, not vice
versa).

tiwapa, '11', was surely adopted from the Hometwoli (Buena
Vista Yokuts) döwāp (Kroeber 1963, p.211). From the same
source Kitanemuk, a Uto-Aztecan language of the Tehachapi
region, also took its term for 11. But our records tell us

nothing of the history of takotia, '12'; all other Chumash
words meaning 12 contain the term for 4, which clearly is not
present here.

When I was a student in the twenties and thirties of this
century it used to be said that those engaged in the search
for cognates in language families would be well advised to
commence their search in the semantic areas of the numerals
and of body part terms. It now seems that theory was clouded
because of the data employed, which was largely of Indoeuropean origin. We have seen above that borrowing of numeral
words from language to language appears to have been very
frequent in aboriginal California. This contrasts with an
Indoeuropean language such as English: here the first ten
numerals have, except for phonological changes, remained
basically unaltered for perhaps five thousand years. I do
not think enough is known about the evaluation of numerical
systems to assign reasons for the difference in the rate of
change. In any event, when Europeans came to California in
the eighteenth and nineteeth century, bringing with them new
linguistic patterns to be imitated, the natives were prepared
by their earlier history to adapt their counting systems to
those of the newcomers. I now look at a few examples of such
change.

Mrs. Yee, my Barbareño informant, gave -- in the 1950's --
for the numerals in that language, native terms for 1 to 12
and for 16. All the others were transliterations of Spanish
words. Thus, for 13, 14, and 15, I received tilesi (Spanish
trece), katolsi (Spanish catorze), and kinsi (Spanish
quince). The strength of the old system is shown by the
survival of the first twelve terms and of 16, which of course
were critical points in the native system.

We have seen that the Cruzeño record shows for 20 and for

100 expressions meaning 'two tens' and 'ten tens'. Recordings of the later 19th and early 20th century show similar constructions in all Chumash languages; such constructions merely illustrate the decay of the native systems.

A somewhat different manifestation of foreign influence is seen in the following example. About 1890 Juan E. Pico, a literate speaker of Ventureño, was engaged by H.W. Henshaw, an investigator for the Bureau of American Ethnology in Washington, D.C., to record Chumash data for him. In a letter of Pico to Henshaw of 1891 a list is given of the Ventureño terms for quantities greater than 100 (Heizer 1955, p.189). The word for 100 here appears as chijipsh. This is surely our old friend, the inherited word for 16, now given a new function. In the native system it expressed the product of the basic four when multiplied by itself; in 1890 it was used as the name of the new basic term 'ten' when multiplied by itself. That is, its function remains unaltered, although its meaning is quite different. The replacement of the native system by the intrusive decimal one in the counting after 'ten' freed the old term for 16 for a new employment; an appropriate one was found for it.

How did the Chumash come to count by fours? I could learn nothing about this from Mrs. Yee. She was born too late to remember any information about it. I was grateful indeed that she knew the language so fluently, a piece of good fortune that could not have been expected in the middle of the 20th century. The amount of ethnographic material she could give me was, in contrast, very scanty. I have seen it stated in the literature on quaternary counting systems that some speakers of such languages could report the practice of holding sticks between the fingers; but I have never heard of that practice among the Chumash, and Mrs. Yee knew nothing of it. If a linguist, equipped with the knowledge available today, could have worked with these Indians a century or more ago, he would surely be able to give us an answer to the question posed in the first sentence of this paragraph. But it was too late, by the 1950's and 60's; the tradition had been lost. It is, I fear, hoping for too much to think that it was already written down in the past and has been overlooked.

6. Cultural Ecology of Mathematics: Ojibway and Inuit Hunters

J. Peter Denny

In this paper I examine mathematical concepts among Ojibway and Inuit hunters with a special purpose in mind: to
discover the origins of mathematical thought in the simplest
of human societies, the hunting band. We will try to understand why very little mathematical thought is needed for the
hunting life, and then ask ourselves what is different in
more complex societies that leads to the development of
mathematics.

The Ojibway and the Inuit (formerly called Eskimos) are
very different racially, culturally, and in terms of the
natural ecology of their hunting territories. The Ojibway
belong to the main migration of Amerindians from Asia taking
place about 40,000 years ago. They are related to many other
Algonquian tribes such as Micmac, Montagnais, Cree, and
Blackfoot. They hunted in the boreal forest of northern
Ontario, centred above Lakes Huron and Superior. The Inuit
belong to a much more recent migration from Asia taking place
about 6,000 years ago; they hunted on the open tundra and
sea-ice of the Arctic Ocean coastline and islands. Despite
these differences we will see that their mathematical concepts are much the same, therefore I take them to be reasonably representative of hunting societies.

The distinctive thing about hunting economies is that one
gains a living from wild plants and wild animals, in contrast
to the domesticated plants and animals of agricultural economies. Furthermore, only human energy is used, not that of
large domesticated animals or man-made engines as in

agricultural and industrial societies. The dependence of the
hunter upon wild plants and animals leads to two crucial features in his pattern of living. First of all, he only alters
the environment to a small degree and must for the most part
adapt to its natural conditions. In contrast to this, agricultural and industrial societies alter the environment to
increasing degrees and strive hard to make the environment
fit their needs. The second feature arises as a consequence
of the first. Since the technology needed for a small degree
of alteration of the environment is itself restricted, any
adult knows the whole repertoire. Consequently, there need
be no specialization of occupation -- anyone can kill an animal, butcher it, and cook it; anyone can cut wood and bark
from trees, shape them into a canoe, and paddle it. Because
tasks are not shared among specialists anybody can support
himself by his own efforts without reliance on anyone else,
although cooperation with others will normally increase success. In contrast, as the degree of alteration of the environment grows in agricultural societies, the range of skills
multiplies, and tasks must be divided among specialists with
consequent dependence on others in gaining one's living. We
will see many points at which the hunter's mathematical concepts are affected by these two features: first, adapting to
a little-altered natural environment, and, second, performing
all tasks oneself independently of other people. In general,
we will note that because of these features the hunter needs
only a small amount of mathematics, but that as the features
change to their opposites in agricultural and industrial
societies the need for mathematics grows. These opposites, a
high degree of alteration of the environment and division of
work among specialists, require mathematical thinking. In
this vein, I will try to show that mathematical thought is

not inevitable or innate in human beings, but arises from specific conditions in recent human history.

Before we can discuss mathematics in hunting societies, we have to lay aside certain misconceptions which often arise about hunters. First, there are no biological differences among hunters, agriculturalists, and industrialists -- all are members of the same human species which became fully-evolved about 50,000 years ago. Because of this there are no differences in thought capacity or language development between hunters and people in complex societies -- abstract thought is just as highly developed and language is equally complex and flexible. To put it simply, there is no such thing as primitive thought or primitive language. A second misconception is that hunters work harder than farmers and industrialists -- in fact, the opposite is the case: the hunter's work week has been shown to be about 20 hours, whereas wage-work plus house-work runs about 60 hours a week for our industrial society. The relationship involved is that the more the environment must be altered to provide sustenance, the more work must be done -- this is only par-tially alleviated by the harnessing of domestic animals and powered machines as helpers. As a consequence, hunters have ample time for recreation, including intellectual activities such as the elaboration of a corpus of myths. I review these misconceptions so that no one will suppose that mathematical thinking is underdeveloped in hunting societies through lack of capacity for abstract thought or lack of time for, or interest in, intellectual activities. The reason is solely that mathematics has little use in the hunting life.

In looking at mathematical thought we are considering one specialized kind of abstract thought. Abstract thinking in general is equally well-developed in all societies and is equally well-expressed in all human languages. Indeed, if there is any attenuation in abstract language it occurs in the languages of industrial nations because so many abstractions are siphoned off into specialized languages such as logic and mathematics. The languages of hunting societies have excellent representation of abstract ideas such as logical quantifiers, e.g., Inuktitut (the Inuit language) uses the logical quantifier 'only' to express what English conveys with a metaphoric use of <u>around</u> in cases such as <u>he is just walking around</u>, i.e., concentration on only one action (Denny 1981). Thus, <u>pisu-tuaq-tuq</u>, 'walk-only.action-he.does', signifies 'he is doing only one action, which is walking'. Therefore we can recognize that what is underdeveloped is that specialized aspect of abstract thought which we call mathematics. Those particular abstractions, developed in the more complex agricultural societies and in industrial society, are mostly concerned with number and measurement including special applications to space.

COUNTING

An understanding of the role of counting in human life can be gained from a story concerning the famous law-suit of the East Cree against the James Bay project of Hydro-Québec. A lawyer for the developers was questioning a Cree hunter appearing as a witness, attempting to show that the hunter did not have an intimate knowledge of his hunting territory. He asked, "How many rivers are there in your territory?" -- the hunter did not know. The lawyer turned in triumph to the judge believing his point to be made. What he, and probably the judge too, did not understand was that this ignorance of the number of rivers was evidence for the opposite conclusion, that the hunter had a particularly intimate knowledge

of his territory. The hunter knew every river in his territory individually and therefore had no need to know how many
there were. Indeed, he would know every stretch of each
river as an individual thing and therefore have no need to
know in numerical terms how long the rivers were. The point
of the story is that we count things when we are ignorant of
their individual identity -- this can arise when we don't
have enough experience of the objects, when there are too
many of them to know individually, or when they are all the
same, none of which conditions obtain very often for a
hunter. If he has several knives they will be known individually by their different sizes, shapes, and specialized
uses. If he has several pairs of moccasins they will be worn
to different degrees, having been made at different times,
and may be of different materials and design. On the other
hand, domestic articles in industrial society often cannot be
individualized because they are identical -- all one can do
is count the glasses or bowls, the pairs of underwear or the
white shirts. When enumeration is appropriate for a hunting
society small numbers will suffice -- the number of fish of
one species caught on one occasion, the number of beaver in a
particular lodge, or the number of canoes in one party will
never rise very high. In industrial society, one often has a
need to account for thousands and millions of items which are
identical.

The much smaller utility of counting in hunting economies
accounts, I believe, for the greater variability and complexity of the encoding of numbers in the natural languages
spoken by hunters, in contrast to the more compact and homogeneous systems found in the languages of the industrial
world. Where English has separate lexical items for 1-10,
Inuktitut (Aivilingmiut dialect) has them for 1-5 and for 10:

1	atausiq	6	arviniliit
2	marruuk	7	marruungnik arviniliit
3	pingasut	8	pingasunik arviniliit
4	sitamat	9	qulingiluaqtut
5	tallimat	10	qulit

Most importantly, there are heterogeneous ways of conceiving
of the numbers that are not given basic terms, none of them
used in European number words. The term arviniliit, 'those
at the edge of the right hand', appears in the group 6, 7,
and 8, and refers to the fact that these are counted with
little finger, ring finger, and middle finger in traditional
finger counting. This word, unmodified, is used for the
first one in the group, 6. Modified by the word for 2,
marruungnik arviniliit, 'the second one at the edge of the
right hand', it signifies 7, and modified by the word for 3,
it signifies 8. The principle involved is counting within a
subgroup. Another principle is seen in the word for 9,
qulingiluaqtut, 'almost ten', -- the notion is that of
approaching the base unit as a limit.

Similar structures are seen in the Ojibway number words
(given in Odawa dialect):

1	bezhig	6	ningodwaaswi
2	niizh	7	niizhwaaswi
3	niswi	8	niswaaswi
4	niiwin	9	zhaangaswi
5	naanan	10	midaaswi

The words for 1-5 are basic, whereas the second five are
identified as a group by the suffix -aaswi. Counting within
this group is done for 6, 7, and 8 -- 6 is identified by the

alternate root for 'one', ningodw-, 7 by niizhw-,'two', and 8 by nisw-, 'three'. How the roots zhaang- and mid- came to identify the last two members of the group is not known.

For higher numbers native number words also show greater heterogeneity and less compactness than European number words. Two principles familiar in the latter are used -- addition and counting base units (either tens or twenties):

		Inuktitut	Ojibway
Addition	13	qulidlu pingasudlu	midaaswi ashi niswi
		'ten-and three-and'	'ten and three'
Counting	30	pingasut aggait	nisimidana
base units		'three ten's'	'three ten's'

The two languages diverge for the 100's, Ojibway having a new base unit, 100, expressed by the suffix -aakw, whereas Inuktitut has to add groups of tens. Thus, for example, 300 is rendered in Ojibway as niswaak, 'three hundred', while in Inuktitut it is given by avatit aggaidlu qulidlu aggait, 'twenty ten's and ten ten's'.

For counting thousands both languages invoke a new principle, multiplication:

3000	Ojibway	nising midaaswaak,'three-times ten-hundred's'
	Inuktitut	avatit aggaidlu qulidlu aggait quliiqtaqtugit, 'twenty ten's and ten ten's ten-times'

The concept of multiplication is expressed by the Ojibway particle suffix -ing, 'times', and by the Inuktitut verb suffix -iqtaq-, 'times', plus an appropriate verb inflection.

In interpreting European number words the counting of base groups, e.g., thirty, is often viewed as multiplication. However, the existence of a separate structure which clearly is multiplication in Ojibway and Inuktitut shows that it is a mistake to consider counting base units, (e.g., thirty as 'three ten's') to be equivalent to multiplication, (e.g., thirty as 'three times ten'). It is the same mistake that we would make if we regarded counting units (e.g., three as 'three one's') to be equivalent to multiplying units (e.g., three as 'three times one'), and we would certainly not make that confusion. The difference is easily seen when both principles are applied in larger number words. For example, in Inuktitut, 1000 is given by qulit aggait quliiqtaqtugit, 'ten ten's ten-times'. In this term, counting groups of base 10 gives 100, qulit aggait, 'ten ten's', and then multiplying that, quliiqtaqtugit, 'ten-times', gives 1000.

Not only is counting base units different from multiplication, but the count and the base word are separately conceived and encoded in language. In Inuktitut qulit expresses a count of 10 and aggait a group of size 10 which might be counted. In Ojibway the root midaasw- is the count of 10 and the suffix -midana is a group of size 10. In Ojibway the count is always expressed by the first morpheme in the word, the root, and the unit being counted by the second morpheme, the suffix (e.g., midaasw-aak, 'ten-hundred's', 1000; nisi-midana, 'three-ten's', 30). This pattern requires crucial adjustments in the form of very high numbers -- when using the multiplicative principle for counting thousands (e.g., nis-ing midaasw-aak, 'three-times ten-hundred's', 3000), roots in additive combinations can be used up to 19 leaving the suffix position available for -ing specifying 'times' as the unit being counted (e.g., midaach-ing ashi zhaangach-ing

midaasw-aak, 'ten-times and nine-times ten-hundred's',
19,000). However, at 20 the suffix position becomes filled
by the morpheme for groups of 10, - midana [shortened to
-dana for 20], niizh-dana, leaving no place for -ing. This
is solved by adding the relative root dach-, 'so many', to
the suffix -ing forming an extra word (e.g., niizh-dana dach-
ing midaasw-aak, 'two-ten's so many-times ten-hundred's',
20,000) so that all three words give a count and a unit being
counted. A remarkable extension of this structure will be
described in a subsequent section, in which the suffix for
the unit counted expresses not the size of the unit but the
sort of unit -- -ing for the sort 'times' is a first taste of
this.

In this section on counting, I have argued that counting
is of less utility for hunters because most objects are known
individually, whereas industrial technology yields many
objects which are identical or unknown, and which must therefore be apprehended by counting. Because of the lesser
utility of counting, the number words in the languages of
hunting societies have fewer basic terms, and terms are combined by a greater variety of mathematical relations.

THE HISTORY OF INUIT NUMBERS
    The points made in the previous section about the struc-
ture of number words are clarified when we consider the older
forms separately from developments since these hunting groups
entered into trade with the Europeans. This has been done
for the Inuit numbers by Gérald Noelting and his collabora-
tors (Baillargeon, et al. 1977). The ancient set of numbers
which predates the early 19th century fur trade is thought to
be the following:

1	atausiq
2	maqruuk
3	pingasut
4	sitamat
5	tallimat
6	arvinilik atausirmik
7	arvinilik maqruungnik
8	arvinilik pingasunik
9	arvinilik sitamanik
10	qulit
11	itikkanuuqtuut atausirmik
12	itikkanuuqtuut maqruungnik
13	itikkanuuqtuut pingasunik
14	itikkanuuqtuut sitamanik
15	itikkanuuqtuut tallimanik
16	arviqtangat
17	arvitanganit aqraqtut
18	arvitanganit pingasut
19	arvitanganit sitamat
20	avatit

In this ancient series we see three kinds of number words: 1) independent terms for 1-5, 2) terms for three groups of higher numbers: arvinilik for 6-10, itikkanuuqtuut for 11-15, and arviqtangat for 16-20, and 3) special terms for the completion of a group: tallimat for 5, qulit for 10, and avatit for 20 (none for 15). The principle for building compound words is one we reviewed earlier when examining contemporary Inuit numbers, counting within a group. It is the only one applied in this ancient series -- the other principles we saw in contemporary numbers, such as approach-a limit (e.g., qulinqiluaqtut, 'almost ten', 9) and adding (e.g., <u>qulidlu pingasudlu</u>, 'ten-and three-and', 13) are not found.

This ancient structure seems to be determined by counting on fingers and toes. The word for 5, the completion of the first group, is related to 'arm', that for 10, the completion of the second group, is related to 'top' referring to the upper 10 digits (on the hands), and that for 20, the completion of the fourth group, is related to 'limbs' referring to completion of counting on all four limbs. Similar connections are seen in the words for the groups themselves: that for 6-10 refers to the right hand used for counting these numbers, that for 11-15 refers to the feet, and that for 16-20 to the right foot. The body is thus split first into upper and lower digits, then into left and right digits, yielding four groups within each of which counting from 1 to 5 is done. Noelting describes this mapping of numbers to anatomy as a "prototypical" representation of number because each set of five bodily digits is a prototype for the associated group of five numbers. It is also an example of what Lévi-Strauss (1962) calls "the science of the concrete", in which binary oppositions within one domain, numbers, are linked to binary oppositions in another domain, anatomy: the first ten and second ten numbers are linked to the upper and lower digits, and within each ten, the first five and the second five are linked to the left and right digits. Noelting suggests that counting proceeds from left to right because of the further linkage of left-right to east-west and thereby to the rising and setting of the sun -- left is thus a beginning and right an end point. Structures of this sort are one of the main ways of organizing thought in hunting societies. They provide very strong connections across domains so that one set of ideas is always seen in the context of another set. In this case the sequence of numbers is supported by the context of anatomical structure. In a subsequent section we will see other kinds of contextual linkages for mathematical ideas in these hunting societies.

Beyond the first 20, it appears that there was counting of groups of size 20 for higher numbers (e.g., avatit tallimat, 'twenty's five', 100), and a higher unit of 400, avatimma-riit, 'real twenty', which could also be counted (e.g., avatimmariit maqruuk, '400's two', 800). The aboriginal methods of creating compound numbers appear to be counting within a group and counting number of groups.

Noelting hypothesizes that the other structures emerge in the 19th century due to the beginning of trade with the Euro-peans; these include iteration, approaching a limit, addi-tion, and multiplication. Noelting regards some of these as "figural" representations of number and some as "symbolic" representations. By figural he means concepts that are sup-ported by notions of spatial organization -- the clearest case is iterative formation in which one number repeated yields another number. Thus, for some Inuit tribes 6 is pingasuujuqtut, 'three repeated', 8 is sitamaujuqtut, 'four repeated', and 10 is tallimaujuqtut, 'five repeated'. These may have arisen from some spatial display of the objects counted, or tallies of them on counting sticks. The notion of approaching a limit seems to be ancillary to this one: 7 is represented as sitamaujunngigaqtut, 'not quite eight', and 9 as tallimaujunngigaqtut, 'not quite ten'.

It is the additive principle which Noelting regards as essential to a symbolic representation of number, when taken together with counting groups of base size. The count of groups is always a starting point for adding the same set of unit numbers (e.g., in contemporary Aivilingmiut dialect,

aggait pingasudlu marruuglu, 'ten's three and two', 32).
This is symbolic because each element in the words stands for
a mathematical notion without support from prototypical or
figural context; counting groups of the same size and adding
the same set of units to them employs a structure which is
primarily mathematical. In contrast, separating numbers into
groups which fit human anatomy and only using unit numbers to
count within those groups, as occurs in the ancient Inuit
numbers, gives context a strong role in structuring the number system. Noelting's work suggests that trade with the
Europeans decreased the contextual information that is linked
to number concepts. In subsequent sections, we will see
other ways in which context is linked to mathematical ideas
in these hunting societies, and will again see that the contextualization is weakened by European influence.

The last way of forming compound numbers, multiplication,
isn't considered by Noelting but it appears to be figural, at
least in the sense of depending on organization in time.
When the counting of base groups reaches its limit, multiplying the result may be resorted to (e.g., qulit aggait quliiqtaqtugit, 'ten ten's ten-times', 1000); the morpheme -iqtaq-
refers to doing something so many times, therefore the notion
appears to be repeating 10 times over the action of counting
10 tens.

COUNTING IN CONTEXT

A salient part of our modern conception of mathematics is
that it is pure in the sense of being disconnected from the
world -- it is concerned primarily with mathematical relations themselves and only secondarily with how these describe
other things. This emphasis on the content-free nature of

mathematics is quite recent; although present as a philosophical ideal in Greek times, only in the last few centuries has
it become a part of everyday common-sense. I think this
style of mathematical thought mirrors the style that is
required by industrial technology -- the isolation of crucial
variables and their manipulation independent of context. For
example, this is needed for obtaining heat energy from sub-
stances like coal and oil, and guiding that energy precisely
through the transformations involved in practical heat
engines. The development of pure mathematics concerned with
an exact understanding of selected fundamental structures
follows the same pattern of thought -- key variables must be
isolated from their contexts to be fully understood.

This isolation and control of key variables is needed for
the high degree of alteration of the environment achieved by
industrial society. It is inappropriate, however, for people
in hunting societies who earn their living from a relatively
unaltered natural environment. Adjustment to the natural
environment, so that its wild plants and animals may be
gathered, requires a quite different style of thought. The
hunter must be inclusive in his knowledge of what goes on in
nature. Since he will not be controlling nature, isolated
knowledge of crucial variables would not help, but inclusive
knowledge of the whole pattern of natural processes will be
imperative. The difference is clear in the case of navigation. The navigator in industrial society can derive a
direction from a single selected factor, magnetic north, and
maintain this information in isolation by a gyroscopic compass even if the magnetic field becomes disturbed. The navigator in a hunting society must pay attention to dozens of
factors simultaneously. As Carpenter (1973) describes, the
Inuit hunter traveling by dog-sled through a blizzard must

register changing conditions of snow, ice, wind, temperature, and humidity, all in large complex patterns, which taken as whole structures will successfully specify location. The changes in snow, ice, etc. have to be understood relative to what is usual for that location, in that season. They must be thought of in relation to the pattern of changes observed at the last location and expected at the next one. This high degree of contextualization can only be achieved if there is a general habit of treating information in context rather than in isolation. Therefore, it will not surprise us if this inclusive style of thought shows up in the mathematics of hunting societies, just as the isolating style does in the mathematics of industrial society.

The pattern we will see many times over for the Inuit and Ojibway hunters is that non-mathematical information is closely linked to mathematical information, so that the former is kept salient as the context of the latter. In the previous section we saw how anatomy and spatial organization provided context for the Inuit numbers. Now we will examine context, particularly in the formation of complex number words in which compact expression is given at the same time to numerical value and other information pertinent to what is being counted. A simple way in which the context of counting is signalled in Inuktitut is that the number words are all nouns which have typical noun morphology such as inflections for grammatical number (singular, dual, and plural). This means that a number word designates not just the property of being a certain number, but the larger entity, a set of elements having that property. Thus, <u>pingasut</u>, 'three', which consists of root <u>pingasu-</u> and plural marker <u>-t</u>, designates sets of size three. All numbers above 3 also have the plural marker, e.g., <u>sitama-t</u>, 'four', <u>tadlima-t</u>, 'five', etc.;

marru-uk, 'two', has the dual marker -uk, and atausiq, 'one', has no grammatical number marker indicating it is a singular noun, designating sets with one member. The number words of Inuktitut, then, do not isolate the predicate of numerosity but express it in association with the argument to which it applies, the elements of a set. The plural marker of Inuktitut is sometimes -t and sometimes -it; for the number words a systematic use is made of this difference: whereas pingasu-t means 'three', pingasu-it means 'three groups' -- the unit of enumeration is changed from members of a set to sets themselves. This applies to all numbers: atausiit tuktuit (1 group of cariboo), marruit tuktuit (2 groups of cariboo), etc. In each case the number word designates sets of plural members and gives the number of sets, not the numerical value of the plurality of their members.

Since the number words are nouns designating elements in a set of a given numerosity, other relationships applying to these elements can be expressed using other noun inflections, namely the case endings and the possessive endings. The case endings show a variety of relations of the enumerated elements to other entities in the situation, e.g., using the accusative case -nik to express the material for an action as in marruung-nik sanasimajuq, 'it was made from two of them', or using the dative case -nut to express the instrument used in an action as in marruung-nut sanasimajuq, 'it was made with two of them'. Although these inflections give succinct expression to relations of the enumerated set to other entities, they are also used in striking ways to express relations among mathematical entities themselves. For example, ordinal numbers are formed with the possessive endings to indicate that the enumerated elements "belong" to a series (e.g., pingasu-ngat, 'their three', third). The ordinal

series is generated from its first member, sivulliqpaaq, 'the
first', which refers to foremost position in the series --
sivu, 'front', sivulliq, 'first of two', and sivulliqpaaq,
'first of all'. The second member is tuglia, 'it's next
one', -- the concept is of succession in the series, but the
Inuktitut word uses the possessive ending -a to signal that
the next one "belongs" to the first one. From that point on
the numerical roots are used: pingasu-ngat, 'their 3',
(third), sitamangat, 'their 4', (fourth), etc. The enumerated members of the series are related to the whole series as
evidenced by the plural possessor signalled by the ending of
pingasungat, 'third', -- if the relation was to just the previous two members of the series a dual possessor would have
been signalled by pingasungak. We have seen that the Inuktitut ordinals are built up from the concepts of foremost
position, next position, and then enumerated positions, in a
series. The distinctive aspect of the Inuit ordinals is that
the relation of each ordinal to the series is explicitly
encoded by the possessive endings, which I have interpreted
to be a special emphasis on the context which is necessary
for an ordinal number, its series.

The context of counting is also signalled in Inuktitut by
a large number of suffixes which can be added to number roots
to form complex number words. I will describe four of these
which will show the variety of contextual information
involved.

First, if events rather than things are being counted, the
verb suffix -iqtaq-, 'do so many times', can be added to the
number root, e.g., pingasu-iqtaq-tuq, 'he did it three
times', -- such words are sometimes termed iterative numerals. Note that since events are being counted, a verb is
formed which refers to the events and allows inflection for

the participants in them, e.g., <u>pingasu-iqtaq-para</u> would mean
'I did it to him three times'. This Inuktitut number word
gives inclusive expression to numerosity, the event nature of
the units being counted, and the participants in the event.

A second context-specifying suffix is <u>-unaaqtiq-</u> which
forms distributive numerals, those describing repeated occurrences of sets of the same size, e.g., <u>marru-unaaqtiq-ȶugik</u>,
'two at a time'. This word provides recognition that the
distribution of the equal-sized sets usually takes place as a
process over time since it means literally 'they being gradually made to be two'. An example of its everyday use would
be: <u>marruunaaqtiqȶugit gaikkit</u>, 'give them two at a time'.
Static states of distribution of equal-sized sets can be seen
as arising out of processes of distribution, so they are
encoded as result states with the additional suffix <u>-sima-</u>:
<u>marru-unaaqtiq-sima-jut</u>, 'they are in two's'. Although this
is a common way of expressing distributive number the amount
of context that is indicated, gradual making of two's, makes
it inevitable that other expressions will be used when the
context varies. It appears that distribution at one time is
better expressed by <u>-unaaq-</u> without the succeeding suffix
<u>-tiq-</u>, 'gradual': <u>tadlima-unaaq-pait</u>, 'he makes them into
4's'. Another context, the regular occurrence of a certain-
sized distribution, can be signalled by <u>-uqattaq-</u>, 'often
be': <u>marru-uqattaq-ȶugik</u> 'they being regularly two'. These
forms illustrate the rich and varied contexts which are
included in the expression of distributive number.

A third kind of context is the usualness of a given numerosity, expressed by the suffix <u>-usuuq-</u>, 'usually be', [some-
times <u>-ujuuq-</u>]: fingers are usually 5, <u>tadlima-usuut</u>, 'ones
that are usually five', table legs 4, <u>sitama-ujuut</u>, 'ones
that are usually four', etc.

A fourth kind of context is objects having a certain number of parts which can be indicated by the suffix -lik, 'one
which has', e.g., sitama-lik, 'the one with four', such as a
four-engined airplane. This suffix also forms words for the
numerals: sitamalik, 'the numeral 3', possibly because early
exposure to the Arabic numerals was through playing cards
which had the given number of spades, etc., depicted on them.

In all these examples we have seen that Inuit counting is
often expressed using complex number words which also indicate some of the context in which counting takes place. I
believe that this is part of a general cognitive strategy
used by hunting peoples to include together in a single
representation a lot of information from the situation being
dealt with.

THE CONTEXT OF COUNTING IN OJIBWAY

To support the generality of the position just taken, it
is important to show the same phenomenon in another unrelated
hunting group. For this purpose we will examine the structure of Ojibway number words, looking for evidence of coding
for context. We have already introduced the two-part structure of the Ojibway number word as having this kind of role:
the first part, the root, providing the count, and the second
part, the suffix, indicating the unit being counted (e.g.,
nis-imidana, 'three-ten's'; nis-ing, 'three-times'). In
fact, there is a whole system of suffixes, called numeral
classifiers, for indicating what sort of thing is being
counted. These include sets of size 10, -midana, and 100,
-aakw, as well as sets of unspecified size signalled by
-ewaan, 'set [including pairs, teams, etc.]' (e.g., niizh-
danaw-ewaan, 'twenty sets'). Human groupings such as work
groups, families, etc., can be indicated by -oode (e.g.,

nis-oode, 'three families').

The most important group of numeral classifiers are those indicating the properties of objects used in the manual technology of traditional Ojibway life. These classify objects according to their hardness, flexibility, and dimensionality, i.e., relative size in the three spatial dimensions, all features which affect their manipulability. The classifier -aabik is used when counting hard inorganic solids such as rock and metal (e.g., midaasw-aabik asiniin, 'ten-hard stones'). Such materials are so hard that they cannot be shaped by bending, but only by chipping and grinding. The other classifiers are for softer organic solids, such as wood, bark, roots, plant fibers, bone, sinew, and hide, which can be shaped by bending and cutting. These are further distinguished by their dimensionality into classes for three-dimensional objects like fruit, two-dimensional objects like bark, and one-dimensional objects like branches and roots. The latter are further distinguished as rigid, like branches, or flexible, like roots. The classifier for three-dimensional objects is -minag (e.g., niizho-minag miinan, 'two-3D blueberries'). That for two-dimensional objects is -eg (e.g., niiw-eg ozhashkwayaanag, 'four-2D muskrat skins'). For one-dimensional rigid objects it is -aatig (e.g., ningodwaasw-aatig misan, 'six-1D.rigid pieces of firewood), and for flexible one-dimensional objects it is -aabiig (e.g., naanwaabiig wadabiin, 'five-1D.flexible roots'). The direct relevance of these classes to the traditional manual technology can be seen when we consider the manufacture of the most complex and subtly-designed artifact, the canoe. To make a canoe, one-dimensional rigid pieces of wood are cut and bent into the shape of the frame members. These are then tied together by one-dimensional flexible roots, and finally the

covering is made with two-dimensional sheets of birch-bark.
Similar combinations of materials from these three classes
are involved in making a wigwam or a birch-bark container.
The five classes we have discussed, inorganic solids
(-aabik), organic solids: 3D (-minak), 2D (-eg), 1D.rigid
(-aatig), and 1D.flexible (-aabiig), are all concerned with
qualities of objects which affect their manipulability. They
are, therefore, specified whenever one is counting manipulable objects (e.g., nisw-aabik zagahiganan, 'three-hard
nails'; zhaangaso-minag bikwaakwadoon, 'nine-3D balls';
nishwaasw-eg waabowaanan, 'eight-2D blankets'; naanw-aatig
mandaaminag, 'five-1D.rigid corncobs'; niizhadana dasw-
aabiig ginebigoog, 'twenty so.many-1D.flexible snakes').
However, for objects which cannot be manipulated, no class
can be indicated, so the simple numerical word with no suffix
is used (e.g., niizh makoog, 'two bears'; niswi aanakodoon,
'three clouds'). This also tends to be true of complex
manipulable objects which do not belong saliently to any of
the five classes (e.g., niizh makizinan, 'two shoes').
Besides the five classifiers for manipulable objects Ojibway
has special classifiers for counting the two most important
artifacts made within the traditional economy, the house and
the boat (e.g., niizh-oonag, 'two boats'; niizho-gamig, 'two
houses'). All of these numeral classifiers for concrete
objects ensure that, when counting, expression is also given
to essential aspects of the object counted, especially those
that affect the handling of the object. The manipulation of
objects is one of the most frequent contexts for counting in
a hunting economy where each person constructs by hand his
own clothing, housing, transport, tools, etc.
    Besides indicating the context of counting by suffixes
following the numeral root, certain quite abstract aspects of

context are indicated by changes in the numeral roots themselves. This is achieved in three ways: 1) by the presence
or absence of -w- on the numeral root (e.g., nis-w- in
nis-w-eg, 'three 2D objects', compared to nis- in nis-ing,
'three times', 2) by the use of two roots for the number 1,
bezhighw- and ningodw-, and 3) by reduplicating the first
syllable of the root (e.g., bebezhigw-). The first difference seems to be used to distinguish the counting of concrete
and abstract entities. Most of the time people count concrete objects, so the -w- form of the root is used. This is
true of the independent number particles, e.g., nis-w-i,
'three', niizhwaas-w-i, 'seven', built up apparently from the
number root, the -w-, and a particle final -i. [Nonetheless,
many are irregularly formed, particularly naanan, 'five', and
niiwin, 'four'.] The -w- form of the root is used when the
numeral classifiers for concrete objects are added (e.g.,
nis-w-eg, 'three 2D objects'). However, when abstract entities such as 'times' and 'ten's' are being counted the -w-
does not appear (e.g., nis-ing, 'three-times'; nis-imidana,
'three-ten's'; naan-ing, 'five-times'; naan-imidana, 'five-
ten's'; ningodwaach-ing, 'six-times'; ningodwaas-imidana,
'six-ten's'). The root without -w- also appears in the number verbs which mean 'to be N' presumably because all that is
being expressed is the numerosity as a predicate -- information about what is being counted is supplied independently
in the inflection of the verb (e.g., nis-i, 'he is three';
nis-iwag, 'they [animate] are three'; nis-inoon, 'they
[inanimate] are three'; niizhwaach-inoon, 'they [inanimate]
are seven').

    The presence and absence of -w- on the number root helps
us to trace the historical evolution of the Ojibway numbers.
The suffix -aas ~ -aach which forms the numbers 6-10 requires

a -w- on the preceding number root (e.g., ningod-w-aas-wi,
'six'; ningod-w-aach-ing, 'six times'), which points to the
original concrete meaning attributed to this suffix, e.g.,
that it refers to the fingers on the other hand used in
finger counting. In contrast, the suffix used for 'ten's',
-midana, seems to express an abstract idea, simply sets of
size ten, and accordingly no -w- appears on the root (e.g.,
nis-imidana, 'thirty'). However, the suffix for 100's,
-aakw, seems to be based on a recent concrete idea, namely
the European practice of packaging goods in wooden boxes of
100, experienced by the Ojibway during the fur trade. Consequently, the -w- form of the root is used (e.g., nis-w-aak,
'three hundred'). The concreteness of -aakw is further signalled in the number verbs where it is still accompanied by
the abstract verb final -ad which it requires in its original
meaning 'organic solid' [i.e., wood, bone, etc.] (e.g., nisw-
aakw-ad-oon, 'they are three hundred'). We have seen that
-w- following a number root signifies that something concrete
is being counted, and that this indicator may remain as an
archaic sign of original concrete ideas which were the basis
for abstractions such as '100'.

The second variation in Ojibway number roots which provides information about the context of counting is the use of
two different roots for 'one', bezhigw- and ningodw-. The
latter, ningodw- appears to convey the more basic sense of
one in the number sequence, whereas the former, bezhigw-, is
used to emphasize one in the sense of a single entity. Contexts in which a single entity is salient include counting a
single object by the simple number particle, bezhig, 'one',
describing the numerosity of a single object by a number
verb, bezhigo, 'he is one', and bezhigwan, 'it is one', and
counting a single object having particular concrete qualities

by one of the numeral classifiers for concrete objects (e.g.,
bezhigw-aabik, 'one hard object'; bezhigo-minag, 'one 3D
object'; bezhigw-eg, 'one 2D object'; bezhigw-aatig, 'one 1D
rigid object'; bezhigw-aabiig, 'one 1D flexible object'). In
a variety of other contexts, 'one' appears to be treated as
one in a numbered series, and the root ningodw- is used.
This occurs when multiples are being counted rather than
single units (e.g., ningodw-ewaan, 'one set'; ningodw-aak,
'one hundred'), and when events rather than things are being
counted (e.g., ningod-ing, 'once'). A more clear-cut example
of one in a series is provided by the use of ningodw- to
derive the number root for 6, ningodw-aaswi. As explained
previously, this is literally 1 in the series 6 to 10 counted
on the other hand. One of the most interesting contrasts is
that between the five classifiers for concrete objects used
as materials in the Ojibway technology, which are listed
above taking bezhigw- for 'one', and the two classifiers for
artifacts created by the technology, houses and boats, which
take ningodw- for 'one' (e.g., ningodo-gamig, 'one house';
ningod-oonag, 'one canoe'). The difference involved is a
fundamental one between properties and classes: the five
classifiers for materials identify properties of individual
objects, such as, for -aatig, one-dimensionality and rigid-
ity, so that when using them one can conceive of a single
object in isolation (e.g., bezhigwaatig, 'one 1D.rigid
object'). However, the two classifiers for artifacts do not
refer to properties of objects but to classes to which they
belong, boats for -oonag and houses for -gamig. Consequently, when one object is counted it is correctly conceived
as one in relation to the rest of the class and therefore
ningodw- is the appropriate root (e.g., ningod-oonag, 'one
canoe'). Another contrast shows the flexibility of the

system in indicating either one in isolation by bezhigw- or
one in series by ningodw-. Although bezhigw- is normally
used with the five classifiers for materials, ningodw- can be
substituted if the context saliently involves one in a
series. For example, bezhigwaabiig naabikawaagan would refer
to 'one necklace' but ningodwaabiig naabikawaaganag to 'one
of the necklaces'. Finally, the difference for some of the
simpler words which each forms -- the number verb bezhigo,
'he is one', also means 'he is alone', consonant with the
concept of one single entity expressed by bezhigw-, whereas
the particle ningoding, 'once', also means 'formerly', i.e.,
at a previous time, consonant with ningodw- expressing one in
a series.

The third way in which the number root varies to provide
contextual information is reduplication of the first syllable, e.g., bezhigw- becomes bebezhigw-. Reduplication always
signals some kind of iteration, and in the case of number
roots it is sets of the specified size which are iterated
usually in some distribution mapped into other entities. An
interesting example is found in the word coined for the European horse, bebezhigooganzhii, 'one-hoofed', in which the
reduplicated root bebezhigw- indicates one hoof for each
foot, rather than the double hoofs of the native species such
as deer. By reduplication a whole series of distributive
numerals is formed (e.g., bebezhig, 'one each'; neniizh,
'two each'; neniswi, 'three each', etc.). When the roots for
the iterative numerals are reduplicated an iterative distributive series is obtained (e.g., neningoding, 'once each';
neniizhing, 'twice each'; etc.).

We have seen that variations in the number root provide
information about several quite abstract aspects of what is
being counted: roots with -w- are for counting concrete

objects, and without -w- for abstracts such as events and
sets; the two roots for 'one' emphasize the counting of one
in isolation, bezhigw-, or one in series, ningodw-; and the
reduplicated roots indicate some distribution of repeated
sets of the specified size (e.g., bebezhig, 'one each').

The comparison I have drawn so far is between an inclusive, context-sensitive style of mathematical thought typical
of hunting societies, and a selective isolating style characteristic of industrial society. My argument for these cultural ecological differences is strengthened by the finding
that the context indicators in Ojibway mathematical words are
being lost as the Ojibway gradually change from hunting to
occupations in an industrial economy. The current adult generation has by and large made this shift, and only old people
were trained in their youth for hunting. Today's young and
middle-aged adult speakers restrict the use of the numeral
classifiers for manipulable objects, presumably because they
no longer process these things themselves but purchase the
factory-made object. For example, -minag, the classifier for
three-dimensional objects may be dropped for something like
peas -- instead of the traditional form reported in Baraga's
grammar of 1878, nisimidana daso-minag aninjiimin, 'thirty
thus many-3D peas', speakers prefer nisimidana aninjiiminan,
'thirty peas'. In some cases the classifier is still used to
replace a noun but not together with it, so that the traditional niijo-minag mishiiminag, 'two-3D apples', is replaced
by niij mishiiminag, 'two apples', but niijo-minag, 'two-3D
objects', is still used to refer to the apples without using
the noun.

In this section and the previous one I have shown a variety of ways in which Inuit and Ojibway number words indicate
the context of counting by providing information about what

is being counted, e.g., concrete vs. abstract entities,
things vs. events, etc. I interpret this as evidence of a
general tendency in hunting societies to include context with
any thought, contrasting with the tendency to isolate the
selected thought in industrial society reflected in the lack
of contextual information in European number words.

ARITHMETICAL OPERATIONS

For the ordinary citizen of industrial society (and probably its predecessor, the plow agriculture of Western Europe)
the heart of mathematics is the four arithmetical operations,
the mastery of which is essential for secure employment. In
contrast, the hunter has little or no use for such thinking.
To understand this we have to grasp the gulf between counting
and arithmetic, and to see what sort of tasks, absent from
hunting economies, require arithmetic. Finally, we will see
that arithmetic operations are specializations of more general thought operations which hunters do employ, so that the
Inuit and the Ojibway had no great difficulty formulating
them when trade with the Europeans made them necessary.

Counting, which we have examined in the previous sections,
necessarily involves the objects counted -- there must be
some objects to map on to the number series in order for a
count to be obtained. In contrast, arithmetic operations can
manipulate purely hypothetical numerical values. For hunters
the objects involved in earning a living are always manipulated by the person himself so that counting them may be
useful, but doing arithmetic operations on their numerical
values seldom will be. In more complex societies, where a
servant must do the bidding of a master, the master may not
manipulate the objects but may control the servant's doing
so. Suppose a flock is to be divided into two -- if the

shepherd himself does this he will decide on the assignment
of animals to the two new groups on the basis of his knowledge of the individual animals, their relations to each
other, and to the two proposed herders. No count of the
original flock nor of the two new ones need be made and no
numerical division performed. However, if the master who has
never seen the herds does the division how can he do it other
than numerically? He is told that the flock of 80 animals is
too large, and he orders that it be divided into two flocks
of 40 each. The main condition under which arithmetical
operations become useful is economic action at a distance.
Such conditions do not arise for hunters or for the simpler
forms of agricultural society. The basic factors we have
associated with the need for mathematics, increased alteration of the environment and increased dependence on others to
perform specialized tasks, must have developed to the point
where some people are specialized managers of the man-made
agricultural system who direct the efforts of specialized
workers. Even in societies with quite a lot of economic
specialization, face to face trading of goods and services
will preclude the use of arithmetic.

The basic role of the arithmetic operations is to permit
manipulation of the numerical values of objects as a substitute for the manipulation of the objects themselves. The
kind of economic action at a distance that we have been
describing, involving masters who must do arithmetic to
control the actions of their servants, is one that requires
well-developed arithmetic. However, lesser degrees of development may be observed with other kinds of economic action at
a distance, particularly trading through middlemen. It is
likely that the Ojibway hunters participated to some degree
in trading networks when they were allied to the complex

Mississippian culture in the medieval era, and then continued
this as trade with Europeans developed. On the other hand,
it is unlikely that the Inuit felt even this milder pressure
towards arithmetic prior to European contact. As we look at
their terminology for the arithmetic operations, we will see
that the Ojibway is somewhat more abstract and compact suggesting a longer period of usage, whereas the Inuktitut gives
more concrete expression to the details of arithmetical
problem-solving.

In Ojibway, addition is encoded quite abstractly as conjunction: bezhig miinawaa bezhig, mii niizh, 'one and one,
thus two'. This can be abbreviated to bezhig, bezhig, mii
niizh, 'one, one, thus two'. However, an alternative formulation gives some hint of the concrete process of adding:
bezhig geyaabi bezhig, mii niizh, 'one yet one, thus two' --
the word expressing the additive operation, geyaabi, 'more,
yet', gives the notion of successive combination. In the
Inuktitut formulations, indications of process as well as
conjunction are always present. A mild expression of process
is given in forms conveying 'becoming the sum': 2 + 3 = 5 is
marruug-lu pingasud-lu tadlima-nguq-tut, 'two-and three-and
five-become-they' (two and three makes five), whereas a
stronger expression of process occurs when 'combining the
numbers' is also indicated: marruuglu pingasudlu katittugik
tadlimanguqtut, 'two.and three.and someone.joining.them
they.become.five' (two and three put together makes five).

For subtraction, the Ojibway is again quite abstract:
naanan noondaach niizh, niswi ishkose, 'five lacking two,
three remains', whereas the Inuktitut appears to describe the
process of subtracting with Arabic numerals: arviniliit
marruungnik piiqsi-vi-gi-blugit sitamanguqtut, 'six two
take.away-place-have-one.doing.to.them they.become.four',

(someone having six as a place for taking away two, they
become four) -- if this is so the formulation would have
arisen in recent centuries of trade with Europeans. What we
have seen for addition and subtraction are abstract formulations in Ojibway but representations of process in Inuktitut.

For multiplication, both languages employ the morphology
for counting number of occurrences which we have already
examined. In Ojibway, 5 × 2 = 10 is niizh nenaaning, mii
midaaswi, 'two five.times.each, thus ten'. The word for the
multiplier 5, nenaaning, 'five times each', is a distributive
iterative number containing the suffix -ing for 'times' as
well as reduplicating the root naan-, 'five', to yield the
distributive root nenaan-, 'five each'. The conception is of
taking each member of the multiplicand set (of 2) five times
into the product set (of 10) -- an abstract but very detailed
description of the mathematical relations involved. The
Inuktitut representation is similar: 2 × 3 = 6 is expressed
pingasut marru-iqtaq-ᚆugit arvinilinguqtut, 'three two-times-
one.doing.to.them they.become.six' (three taken twice, makes
six), using the suffix -iqtaq-, 'times'.

Division is the most interesting of the arithmetic operations to study in these hunting cultures because it appears
to be absent until trading introduces it, yet is conceptually
related to a domain which is very highly developed, the sharing of produce which has been caught or gathered. The contrast will show that, as I have earlier asserted, numerical
value becomes important only when dealing with unknown or
indistinguishable elements. Traditional methods of dividing
game preserve the distinct anatomical identity of each share.
Inuit, when sharing a seal, divide it into specified anatomical parts which are the proper share of particular relatives
of the hunter. The seal is never treated as a homogeneous

mass of seal meat whose bulk would have to be assessed
numerically and then divided into parts whose size was determined by arithmetical division. In anatomical division the
structure of the material is followed closely. This concern
for structure shows up in Inuktitut words for parts: avvaq,
'natural part', for parts separated by some structural division, nappaq, 'crosswise part', for parts derived by dividing
across the grain (of muscle fiber, wood fiber, etc.), and
quppaq, 'lengthwise part', for parts got by splitting with
the grain. Compared to structural division, numerical division is a very foreign concept and the words used to convey
it, since trade with Europeans made it important, are singu-
larly vague in what they encode. The notion of dividing a
set into two is expressed by marru-ili-nga-jut, 'two-become-
have-they' (they have become two). The crucial morpheme is
-ili-, 'come to be', -- it conveys the idea of arithmetical
division only when combined with a number root such as
marru-, 'two'; otherwise it expresses more general kinds of
becoming (e.g., taima-ili-nga-jut, 'they have become thus').
Another suffix also meaning 'come to be', -uli-, is also used
(e.g., sitama-uli-nga-jut, 'they have become four', i.e.,
they are divided into four). Using this terminology, numerical division can be expressed as follows: 6 ÷ 2 = 3 is
arviniliit marru-ili-blugit pingasunguqtut, 'six two-come.to.
be-one.doing.to.them they.become.three' (someone making six
into two, they become three, i.e., six divided by two, makes
three). Not only does the notion of division have to be
inferred from morphemes meaning 'come to be', but the set
whose number is given by the root is likewise not specified
in the word itself. Usually sitama-uli-nga-jut will be
interpreted to mean 'they are divided into four', i.e., the
divisor is 4, but in appropriate contexts it can convey 'they

are divided into four's', i.e., the quotient is 4. An example would be pingasut sitamaulingajut, 'three they.have.come. to.be.four', (they are divided into three four's); the independent number word pingasut gives the divisor 3, and the number root sitama- on the verb of division gives the quotient 4. What we have seen for Inuktitut is that numerical division is difficult to express except in a very vague way, because of the high development of physical division into natural parts which does not involve enumeration. Ojibway speakers have noted a similar situation for their language: there is no way to express numerical division because of the emphasis upon dividing into natural shares. It seems likely, however, that some dialects, at least, have developed ways to talk about arithmetical division in recent times.

The position I have taken is that numerical operations are not needed for hunting economies because no one must instruct others in work done in his absence -- there are no occasions which require one to manipulate the numerical value of sets of objects in the absence of the objects themselves. When terms for arithmetical operations are found in the languages of hunters, I attribute them to participation in trading through middlemen with distant partners in trade networks set up by more complex societies. Some of the Ojibway terms seemed abstract enough that they might go back to the medieval era when the Ojibway are thought to have traded on the northern fringe of the complex Mississippian culture. The Inuktitut terms showed considerable representation of concrete aspects of arithmetical processing suggesting a more recent origin in trade with Europeans.

SHAPE AS THE PRECURSOR OF GEOMETRY

We are often told that geometry, as its name suggests,

emerged from the need to measure plots of ground. Certainly such needs are not found among hunters, since ownership of large hunting territories is shared with other members of the band, without imposing man-made boundaries. However, there are other impulses toward recognizing fundamental geometric forms such as circle and triangle. An important one is having a sufficient degree of job specialization so that one person may design an object and another one build it -- it is much easier for the designer to control the builder's actions if the design involves regular geometrical properties. This situation does not arise among hunters who always design and build their own artifacts. Consequently, complex and irregular shapes can be accommodated since there is never a need to communicate them from one person to another. The shapes of a canoe or an igloo are extremely sophisticated but do not have to be analyzed in terms of geometrical properties -- perceptual judgments of length and degree of curvature allow the designer-builder to control his own work. Another aspect of occupational specialization in complex societies which calls for regular geometrical shapes is the making of parts by one worker for assembly by another. A circular hole and a cylindrical rod will fit together without difficulty even if the worker has never seen the maker of either hole or rod. However, for a hunter doing his own building and assembly an irregular-shaped rod can be fitted just as easily to an irregular-shaped hole.

If hunters do not need geometrical concepts what sort of shape concepts do they have? The answer is shape categories, exactly analogous with other basic descriptive categories such as red, hard, sour, and smooth. Psychological study of categories has shown that each one covers a range of variation, that understanding of the category frequently involves

knowing its central or prototypical members, and that the
boundaries between categories are vague and require judgement
in individual cases. This means that a shape category such
as 'round' covers varying kinds of roundness from the circular penny to the slightly flattened tomato, the vertically
elongated apple, and the quite irregular potato -- all fall
into the category 'round'. The circle is one possible prototypical or central member of the category 'round'; also, the
boundary line between 'round' and an adjacent category such
as 'angled' cannot be fully specified but has to be judged in
particular situations.

We can see these categorical properties if we examine two
Ojibway shape concepts, say waawiye-, 'round', and noonim-,
'round and oblong'. Both include a variety of regular and
irregular shapes -- noonim-aa, 'it is round', (or noonim-izi
for grammatically animate objects) could describe the shape
of a cucumber, a long potato, the back of a Volkswagen, or an
egg. All of these are curves which are elongated, although
no geometrical properties are specified. The more regular
shapes can be indicated by a modifier; thus, an ellipse could
be described as weweni-noonimaa, 'it is regularly round and
oblong', although, the geometrical properties of the shape
are not involved -- any kind of regularity would be covered
by weweni-.

The boundary line between the category of elongated curves
designated by noonim- and the category of non-elongated ones
designated by waawiye- is as vague as that between other categories such as 'blue' and 'green', or 'rough' and 'smooth'.
It is left to momentary factors of context to determine
whether a potato is elongated enough to be labelled noonim-
izi, 'it is round and oblong'. Moreover, the boundary line
between these shape categories is very different from that

between geometrical concepts such as circle and ellipse. A potato may be somewhat elongated and still belong to the category described by waawiye-, 'round, non-oblong'; however, the slightest elongation of a circle, so that there are two foci instead of one makes the figure an ellipse. Thus, zero elongation is just one value (although it may be prototypical) of the category non-oblong rounded figure expressed by waawiye-, but is the only value for the geometrical figure circle. Geometrical concepts have precise boundaries usually provided by zero values for some relevant variable, whereas shape categories have vague boundaries often falling towards the middle of some relevant variable.

Another example of this is the contrast between the dimensional size categories, one-dimensional, two-dimensional, and three-dimensional, which were discussed in the section on the context of counting, and the geometrical notions expressed by line, plane, and solid. For the three categories the spatial dimensions of length, width and thickness are divided categorically into small and large values with vague boundaries between them. A one-dimensional object like a stick has large length, and small width and thickness. In contrast, a geometric line has any positive length, and zero width and thickness. A two-dimensional object such as an animal skin has large length and width, and small thickness, whereas the geometric plane has positive length and width, and zero thickness. A three-dimensional object such as an apple has either large values or small values for all three dimensions, whereas a geometric solid has any positive value for all three dimensions. The considerable difference between the two conceptual systems is seen in the fact that objects exemplifying each of the dimensional categories (stick, animal skin, and apple) will all count as geometrical solids.

So far, we have emphasized the role of the second ecological factor discussed in this paper, lack of occupational specialization since each producer is both designer and builder in a hunting economy, in the use of shape categories rather than geometrical figures. However, the first factor, subsistence derived from a relatively unaltered environment, also encourages shape categories. The shapes of the natural world are both irregular and highly variable, so that they cannot be efficiently grasped in geometrical terms. However, shape categories such as 'round and elongated' expressed by Ojibway noonim- cover a wide range of natural objects such as leaves, lakes, and fish, as well as man-made objects such as canoes and wigwams. It is only the man-made world of complex societies in which geometrical shapes are common -- and they are there so that designers can accurately control the work of builders, and so that parts made by one specialist can be used by another. The room in which I write contains almost entirely rectangular shapes so that the notion of a 90^0 angle captures much of what I see. To fully understand how shape categories, in their turn, capture the range of variation in shapes of natural objects, it is important to emphasize their, abstractness. Rounded figures are divided, as we have seen, into more and less elongated ones. The more elongated ones designated by noonim- include all degrees of elongation beyond the vague boundary line with the less elongated ones, including infinite elongation. The infinitely elongated round figures are a sub-category which might be called categorically (not geometrically) cylindrical figures. To see this we can imagine a finitely elongated round object such as an egg, and recognize that if it were infinitely elongated the result would be the curved body and straight sides of a cylindrical object such as a stick; in fact, the curvature of

the body of a stick or log is described in Ojibway with
noonim-. Although the curve of the body of a stick is infinitely elongated, the stick is of course finite, having two
ends. The shape of the ends will usually not be elongated,
i.e., they will be roughly circular, and will be described
using waawiye-. These results show not only the abstractness
of the shape categories but their intelligent use in analyzing compound shapes.

Having introduced ourselves to shape categories in contrast to geometrical figures, let us now look at the whole
system of Ojibway categories. First of all, curved shapes
are emphasized at the expense of angular ones, since curves
are what abound in the natural world of rivers, stones,
sticks, fruits, and hills. We have already examined the
contrast between elongated and non-elongated round shapes.
However, both of these categories refer to shapes about an
axis, and contrast with a separate category expressed by
waag- for shapes involving curvature of the main axis of the
object. All these curves can be exemplified by a rope:
noonimaa, 'it is round and oblong', referring to the cylindrical curve of the body of the rope, and waawiyeyaa, 'it is
round', referring to the circular curve of the end of the
rope, both express curvature around the longitudinal axis of
the rope, whereas waagaa, 'it is curved', refers to the curvature of that axis. This difference between shape around an
axis and shape of the axis is also seen in the neutral shape
categories 'flat' and 'straight'. The term nabag-, 'flat',
indicates absence of curvature around the axis, i.e., not
describable with noonim- or waawiye-, but the term gwayag-,
'straight', indicates absence of curvature of the axis.
Thus, if a stick was described as nabagaa, 'it is flat', it
would lack the curvature of its body describable by noonim-,

but if it was described as gwayagaa, 'it is straight', it
would lack the curvature of the longitudinal axis describable
by waag-.

None of these distinctions are found for angles, where
there is just one category 'angular' expressed by ne-. This
can include the angle of a branch which is the angle of its
longitudinal axis as well as the angles of a square-cut
rawhide thong which are about the axis. Some supplementary
notions, however, are used to assess the degree to which
angularity obtains: giin-, 'sharp', describes sharp edges
and a derivative of it giinakw-, 'pointed', describes sharp
points. These contrast with azhiw-, 'dull', for angles which
have some local roundness or lack of angularity at the apex.

The most remarkable aspect of Ojibway shape categories is
that new ones seem to have been formed to accommodate the new
shapes introduced in the man-made objects brought to North
America by the Europeans. The rounded shape of the wigwam
contrasted dramatically with the many angles of the European
cabin. The sight of the European squaring a timber before
being willing to use it must have been striking and bizarre
to the Ojibway. Cross-cultural psychologists call the world
of man-made angles the "carpentered" environment to focus
upon the greatly heightened alterations made in natural materials within European technology and the role of specialized
craftsmen in maintaining the large repertoire of techniques
needed for these alterations. To be able to conceptualize
the new shapes, the Ojibway developed two new categories for
angles. Both of them are for repeated angles since they
recognized that the predominant shapes in the "carpentered"
environment such as rectangle and triangle all involve
repeated angles. One of the categories is for repeated
right angles recognizing the special importance of the right

angle in European constructions -- it is expressed by
gakakade-. Thus, a rectangle or square can be described as
gakakadeyaa, 'it is repeatedly right-angled'. The other
category expressed by zhashawe- is for repeated non-right
angles, i.e., oblique and acute angles taken together. Thus
the angles in a triangle typical of a low-pitched roof,
oblique at the apex and acute at each base angle, can all be
captured when such a triangle is described as zhashaweyaa,
'it is repeatedly non-right-angled'.

These new shape concepts are still categorical and not
geometrical. To be described as 'repeatedly right-angled' by
gakakade-, the angles need only belong to the category of
angles whose central or prototypical value is 90° -- to be
non-right-angled they must be saliently acute or oblique.
This allows a somewhat off-rectangular shape to be described
as gakakadeyaa, 'it is repeatedly right-angled'. For
instance, the shape seen in some coffins, where the sides
bulge slightly, forming oblique angles close to 180° towards
the middle of each side and causing the four main corners to
become somewhat greater than 90°, can still be described as
right-angled. Indeed the four main angles close to 90° can
be emphasized and the two slight angles in the sides ignored
by describing the figure as newing gakakadeyaa, 'it is right-
angled four times', using the iterative number newing, 'four
times'. This geometrically six-angled figure can still be a
member of the category of repeatedly right-angled shapes.
Naturally, these categories can accommodate any other irregu-
larity in the figure such as curved sides. In fact, a glass
dish with two curved sides joining at two acute angles is
described as niizhing zhashaweyaa, 'it is non-right-angled
twice'. In addition, the categorical nature of zhashaweyya,
'it is repeatedly non-right-angled', allows it to cover a

right-angled triangle, since two of the three angles are
not right angles.

Before leaving the Ojibway shape categories, it is worthwhile to note that context is often specified for them just
as it was for numbers, using many of the same terms for
classes of materials involved in the manual technology. For
example, a round three-dimensional object such as a ball is
describable as <u>waawiye-minag-ad</u>, 'it 3D is round', a square
two-dimensional object such as a blanket as <u>gakakade-yiig-ad</u>,
'it 2D is repeatedly right-angled', and a round and elongated
one-dimensional object such as a rope as <u>noonim-aabiigad</u>, 'it
1D.flexible is cylindrical'.

Inuit shape categories show the same basic properties as
the Ojibway ones. The shape category 'round' expressed by
<u>angmaluqtuq</u>, 'it is round', allows varying degrees of round-
ness which may be expressed by modifying suffixes: the perfect roundness of a circle is describable as <u>angmalu-rik-tuk</u>,
'it is perfectly round', and a quite irregular round shape is
describable as <u>angmalur-lak-tuq</u>, 'it is somewhat round'.
Some new words seem to have been developed for the European
shapes, for instance, <u>kippaariktuq</u>, 'it is square', appears
to come from the verb <u>kipijuq</u>, 'he cuts'. However, the whole
system of Inuktitut shape words has not yet been studied.

We have seen that the shapes of objects are conceptualized
by Ojibway and Inuit hunters in terms of shape categories.
These allow for the considerable variation within a category
which is typical of objects in the natural environment. They
are also suitable for the variable shapes of arrows, knives,
toboggans, and moccasins which are both designed and made by
the individual hunter. In contrast, geometric shapes arise
in cultures where specialization of occupation separates
designing from building, or separates the maker of parts from

the assembler of the finished artifact.

MEASUREMENT

In hunting societies such as the Ojibway and Inuit, measurements are frequently made, but are of quite a different sort than those made in industrial society. The measurements are context-sensitive rather than objective. Where industrial measurements are strictly uninfluenced by factors in the measuring task and only reflect the standard unit, the hunter's measurements are sensitive to many factors in the situation at hand. For instance, hunters measure using the size of their own arms and hands as units making only occasional adjustments for any unusual size of these body parts. In the same vein, the size of a container may be determined by fixing a main dimension by perceptual judgment rather than measurement -- a size is picked which looks suitable for the function of the object. This distance is then defined as a unit of measurement for determining the size of other parts of the object, but once the construction is finished the unit is never again used. This sort of measurement is appropriate when someone designs and builds an object himself rather than the work being divided among persons with specialized skills. In contrast, the specialized jobs performed by workers in the industrial world are only able to be coordinated by objective measurements which are free of the particular context of the individual worker's activities. Since designer and builder may never share any work situation, the designer's ideas must be conveyed in context-free measurements, i.e., those that follow standardized objective units.

Measurements among hunters are not only context-sensitive but are also closely coordinated with perceptual judgments. It is probably fair to say that measurement is used when

perceptual judgement might fail. Shape factors such as
straightness and rate of curvature are typically judged by
eye and not measured. However, size relationships such as
having equal sides cannot be judged accurately enough and are
therefore measured. The use of judgment assisted by measurement is an efficient tactic for the individual designer-
builder of hunting societies. However, in complex societies,
perceptual judgment cannot be relied upon, once occupational
specialization takes hold, separating the mind of the
designer and the minds of the various builders.

We will look at these basic features of hunter measurement
in more detail later on, but first we will consider more
broadly the range of things measured in hunting societies.
Measurements are made of linear distance, amount measured by
containersful, and time. Domains of measurement which are
important in industrial society but not in hunting societies
include area, volume, weight, temperature, and monetary
value. Area and volume are unimportant for hunters because
they are secondary values derived from linear distances which
are only needed when the object measured is considered out of
context -- if I am going to use a container I did not build,
I may need a measurement of its volume. However, if I have
designed and built the container myself, the basic linear
measurements of the design will have given me an adequate
appreciation of its volume without calculating this as a separate measure. Weight and temperature can be appreciated
adequately by perceptual judgment, especially since the
person needing the information is always present in the
situation and is never acting at a distance. Another factor
encouraging the measurement of these two variables in industrial society is the high degree of alteration of the natural
environment -- some industrial processes involve weights and

temperatures so great or so small as to go beyond the limits
of perceptual sensitivity. Monetary value is, naturally,
only needed when there is trading at a distance -- in hunting
societies trading is deeply embedded in face to face social
interactions, usually with family members and others in close
personal relations. Reciprocal social obligations completely
outweigh any possible abstract monetary value of the goods
exchanged. It is only when strangers exchange goods that
monetary value is needed to regulate the exchange.

Turning to the domains in which hunters do carry out measurements, let us look first at time. Hallowell (1942), writ-
ting about Ojibway measurement, emphasized the "processual"
nature of their techniques. Time is assessed by the interval
taken up by some characteristic activity, either of humans,
e.g., the number of "sleeps" made in a journey, or of some
natural process such as the changing position of the sun in
the sky. The main time units are those provided by recurrent
changes in the environment such as the reoccurrence of winter
to define a year, reoccurrence of the moon's phases to define
a month, and reoccurrence of light and dark to define a day.
In Ojibway, the most important of these units, the day, is
expressed by a suffix on the numeral root, -gon (e.g.,
ningodo-gon, 'one day'), the same construction we saw earlier
for identifying other units of enumeration such as classes of
objects (e.g., ningod-oonag, 'one boat') and occurrences
(e.g., ningod-ing, 'one time'). The other time units are
expressed by using prefix forms of the numerals placed before
a noun expressing the time unit (e.g., ningo--giizis, 'one
month'; ningo--biboon, 'one winter', i.e., one year). Daytimes and nights may sometimes also be counted in contrast to
the 24-hour day (e.g., niizho--giizhig, 'two daytimes';
niizho--dibik, 'two nights'). An important unit based on

human activity is the time taken to smoke a pipe, ningodopwa-aagan, 'one pipesmoke'. Days, months, and years are also the main time units for the Inuit, expressed by nouns, inflected for singular, dual and plural number to agree with the inflection for size of set on the number word discussed earlier (e.g., atausiq ubluq, 'one day'; marruuk ubluuk, 'two days'; pingasut ublut, 'three days'; pingasut taqqiit, 'three months'; pingasut ukkiut, 'three years').

A special feature of time measurement among hunters is its use to assess distance travelled. Rather than using a standard measure of linear distance such as the mile, a temporal unit such as 'sleeps' or 'nights' is preferred. This tactic has the effect of providing context-sensitivity for the measurement, because the amount of time taken to make a journey is sensitive to the terrain, weather, mode of travel, provisions needed, and other aspects of covering the distance involved, in a way that a measure of linear distance itself would not. This is an example of the inclusive style of thinking in hunting societies discussed previously -- many aspects of context are included with the main information.

A similar concern for specifying the context of measurement in accordance with an inclusive style of thought can be seen for containersful measures such as handsful. In Inuktitut the relatively neutral atausiq aggak tatadlugu apummik, 'one hand full of snow', is often replaced by variants indicating more context:

1) atausiq aggait tatadlugu apummik, 'your one hand full of snow';

2) atausiq aggait tataguk apummik, 'fill your one hand with snow';

3) atausiq aggait tatattuq apummik, 'your one hand is full of snow';

4) igluinnaq aggak tatadlugu apummik, 'a single hand full of
snow';

5) igluktuq aggatit tatadlugit apummik, 'both your hands full
of snow';

6) katiglugik aggatit tatagit apummik, 'fill your hands
together with snow'.

The impression one gets from these many variants is that
there is no conventional objective concept of 'one handful',
but that the inclusive style of thinking requires its formulation within some context. This inclusive style we have
previously related to a low degree of alteration of the
natural environment, and conversely the isolating style of
Western thought, including objective measurements, we have
related to the necessity to isolate single variables when
altering the environment to a high degree.

Containersful are such important measurements for hunters
that they are essentially an open class -- in Inuktitut it
seems that any noun for a container may serve as a measure.
If so it will appear in the singular (e.g., pingasut tituqvik
tatadlugu imarmik, 'three cup [SING.] full of water'), to
distinguish it from the container itself (pingasut tituqvit,
'three cups'). If the specified number of containers are
physically present when they are acting as measures, then the
noun stays in the plural (e.g., pingasut qijurqutit qiqit,
'three boxes of nails'). Similarly in Ojibway almost any
noun for a container can define a containerful. Not only is
the measure noun used in the singular, but the prefix form of
the number is used, not the independent number word, and
'one' is expressed with ningodw- [prefix form ningo-], 'one
in series', rather than bezhigw-, 'one alone' (e.g., ningo-
-makak, 'one boxful', rather than bezhig makak, 'one box';
niizho--makak, 'two boxfuls', rather than niizh makakoon,

'two boxes'). For higher numbers the prefix position may have to be filled by daso-, 'so many', (e.g., nisimidana daso--makak, 'thirty boxfuls', contrasting with nisimidana makakoon, 'thirty boxes'). Besides the main container measures formed from nouns, Ojibway has three general container measures expressed as suffixes on the number roots: -sag, 'rigid containersful', -ooshkin, 'flexible containersful', and -wan, 'backpacksful', (e.g., ningodosag, 'one rigid containerful', probably originally made from bark on a wood frame; niizhooshkin, 'two bagfuls', originally any of the bags woven from bark and grass; nisowan, 'three packfuls', perhaps originally wrappings of animal hide).

Undoubtedly the most developed measurement techniques found among the Inuit and Ojibway hunters are those for linear distance. These are involved in the construction of the most complex artifact, the boat. Kayak-building and canoe-building both exemplify an interplay of perceptual judgment, measurement using standard units, and measurement of proportion which will give us the best understanding of the role of measurement in hunting life. The most frequently employed measures are body part lengths. There is an enormous variety of these but they are best understood as families of variable units. Thus there are various hand and finger widths, various hand spans, various fore-arm lengths, various single-arm spans, and various double-arm spans. For instance, Zimmerly (1979) observed the following variety of handwidths used by a kayak-builder: 1) fist width from index finger knuckle to little finger knuckle, 2) fist and thumb width from outstretched thumb tip to little finger knuckle, 3) three-finger width, and 4) little finger and half of ring finger width. A family of fore-arm measurements is produced by varying the position of the hand at which the unit ends. Taylor (1980)

reports the following variants for a Cree canoe-builder (the
Cree are closely related to the Ojibway): 1) elbow to tip of
middle finger, 2) elbow to tip of thumb, 3) elbow to middle
joint of thumb, and 4) elbow to base of thumb. The single-
arm span from armpit to finger tip may be varied also by the
part of the body, rather than the hand, at which it ends: 1)
tip of left fingers to tip of nose facing opposite, 2) tip
of left fingers to right nipple, and 3) tip of left fingers
to right shoulder, as well as varying the part of the hand,
e.g., armpit to first joint of fingers (gripped around
object). We can think of each family of measures, handwidths, hand spans, fore-arm lengths, single-arm spans, and
double-arm spans, as setting up an order of measurement
within which the variants determine more fine-grained units.
Variation in unit is also obtained by combining different
units from one family, e.g., the beam of a canoe measured by
two different fore-arm lengths, elbow to tip of middle finger
plus elbow to tip of thumb, or combining units from different
families, e.g., the length of a kayak gunwale measured by a
double-arm span from index finger tip to thumb tip plus a
fist width from index finger knuckle to little finger
knuckle.

Obviously these techniques provide all the variety in
linear distance needed for measuring artifacts made in manual
technology. What we need to know is why they are an appropriate system of measurement for tasks such as boat-building
in a hunting society. The kayak or canoe maker is constructing a traditional artifact whose design is agreed upon by the
members of his tribe, and which alters only slowly to meet
changing ecological and functional requirements. He knows
some variations in its design to meet different local conditions, and perhaps different users besides himself. He was

trained by apprenticeship with one or two makers of the previous generation, probably close relatives, and all his
knowledge has to be stored in his own memory, without the aid
of external memories such as boat plans. Under these conditions the body part measures have certain advantages. Since
they are based upon the builder's own body they automatically
relate the proportions of the craft to his own limbs which he
will employ while both building and using the boat. Since
the various body part measures involve qualitative differences in the position of body parts they are much easier to
remember than abstract measurement units -- thus, the relatively unchanging design of the artifact can be remembered
from year to year. Also, the use of body parts ensures that
the standards are available at the work site without having
to carry rulers, measuring tapes or any other external standard as part of the tool kit -- this is an important advantage for the migratory life of hunters who must keep their
tool kit as small as possible. Finally, the varying units
within families make it easy to slightly change size to
accommodate different conditions and users. The one disad-
vantage which we imagine for the body part measures, their
lack of objective precision, never applies to the hunter
because he cannot be misguided by using his own body parts to
control his own work, and his measurements will never be used
by other workers. Even where there are helpers, as is often
the case, they are present in the same situation and able to
observe the whole project -- furthermore, it has been noted
that the helpers take on a portion of the work which will
involve minimal interdependence with the other builders.
Precise and objective measurement is needed when a task is
fragmented into stages done at different times and places by
specialized workers who could not otherwise coordinate their

efforts.

Body part measurements are not the only way in which the Inuit or Ojibway boat-builder guides his work. Considerable use is made of perceptual judgment, especially to determine curvature. Curved parts of the kayak are cut by eye without measurements. The gradual tapering in size of the gunwales is also judged by eye, as is the curved line of the gunwales which this taper has an important role in determining. However, certain gradual curves may have to be measured -- for instance, Zimmerly describes how the small reverse sheer in a kayak (the curve upward from ends to middle) is measured in fingerwidths. Likewise, many alignments such as straightness of the keel may be judged perceptually, but others that are less perceptible such as equal distances on two sides may have to be measured. The essential relationship seems to be to use measurement only when perceptual judgement would not be a sufficient guide.

Although the body part measures and perceptual judgments establish the original linear sizes, measuring sticks are often used to temporarily store a measurement or to reproduce the same measurement in another part of the construction. For example, body part measures set out on one gunwale can be marked on a measuring stick for later transfer to the other gunwale. Similarly, a distance set up on one side can be measured with a stick and then the other side can be adjusted to the same size. Such temporary measures can be counted in Ojibway (e.g., niso--dibahigan, 'three measures'). Another way to temporarily store measurements is by marks, e.g., for the position of ribs and thwarts along a gunwale. Such marks can be given ordinal numbers in Ojibway (e.g., eko--niizhing beshibiihigaadeg, 'where the 2nd mark is').

Besides perceptual judgment and body part measurement,

linear distance is also assessed by proportion. For example, a measured distance may be halved by matching it in a length of string and then doubling the string to produce the half distance. Similarly, the mid-point of a piece of material may be dound by folding it in half.

In our examination of measurement we have seen that the work situation in which it occurs is very different for hunters than for members of complex societies. Typically the hunter designs and builds an artifact for use by himself or a family member. A lot of the work can be guided by perceptual judgments, so that the main role of measurement is to guide the work when perceptual judgment would not be accurate enough. Since the work is done in only one situation the measurements are sensitive to many aspects of that one context. In complex societies, manufacture is divided among specialists who work in separate situations so that context-free standard units of measurement are needed to coordinate their work.

SUMMARY

We have looked at four areas of mathematical thinking in the hunting and gathering societies of the Inuit and the Ojibway, counting, arithmetic operations, shape categories as precursors of geometry, and measurement. The main conclusion that can be drawn is that mathematical thinking is a supplement to more basic information processes. When humans are unable to guide their actions by ordinary perceptions and concepts they resort to mathematical ideas. In the case of counting, enumeration serves as a way of apprehending objects which cannot be perceptually or conceptually identified. Such conditions arise rarely for hunters, since their relatively unchanging environment and small number of man-made artifacts will normally allow things to be perceived and conceptualized as individual objects. Therefore, occasions for counting are few and mostly restricted to lower numbers. Similarly, we saw that measurement was used only when perceptual judgment would not do the job. The shape categories we examined are examples of ordinary conceptual processes that are used instead of mathematical ideas -- they have the same properties as other concepts such as color and texture, and lack the mathematical properties of the geometric notions which are their mathematical counterparts.

Another major conclusion was that the mathematical ideas of hunting peoples are very context-sensitive. This was interpreted as a part of a general cognitive strategy of inclusive thinking -- a lot of background information is integrated with the main thought in any thinking operation. This strategy ensures that all relevant aspects of the environment are appreciated, as a guide to action in a wild environment which cannot be much altered or controlled by humans. For counting, context was often specified by suffixes on the number words indicating the kind of units being counted. In addition, the number concepts were structured to fit the anatomical context used in counting by means of the bodily digits. For measurement, context-sensitivity was achieved by measurement units that varied from task to task and person to person.

The main impetus for the further development of mathematical ideas in agricultural and industrial societies was seen to be the greater efforts made to alter the environment creating an increasingly man-made world. To manipulate nature it is necessary to isolate crucial variables, requiring an isolating or selective style of thought in which particular properties are abstracted from context. The abstract, idealized, and standardized notions of number, geometry, and measurement in Western mathematics are all examples of this approach. A secondary effect of efforts to alter the environment is that the expanding repertoire of technological skills cannot be mastered by each individual, leading gradually to more and more specialized occupations. As work is divided among specialists, mathematics plays a crucial role in coordinating their efforts. Identical parts made by one specialist must be counted to be traded to another specialist, regular geometric shapes enable an assembler to use parts he didn't make, and measurement of monetary value allows trading of materials, parts and artifacts among specialists who do not have any important social relations to regulate exchange. In particular, we attributed the importance of arithmetical operations to economic action at a distance among people with specialized occupations -- situations where someone in a managerial role must control objects that he does not interact with. He can understand sets of such absent objects by their numerical values and act upon them by altering the numerical values through arithmetic operations, such as dividing their numerical value to control a distribution of objects which will be carried out by others.

ACKNOWLEDGEMENTS

I am grateful to the many native people who helped me learn about their mathematical concepts. For the present paper I have depended especially upon information gathered by Lorraine Odjig of Wikwemikong, Ontario. I also want to thank the many linguists who have tried to make me less naive about the complexities of Ojibway and Inuktitut. Portions of this work have been supported by the Department of Indian Affairs, the Ontario Department of Education, and the Social Sciences and Humanities Research Council of Canada.

7. Tallies and the Ritual Use of Number in Ojibway Pictography

Michael P. Closs

THE OJIBWAY, BIRCHBARK AND TALLY RECORDS

The Ojibway are Algonkian-speakers who in historical times have occupied a widespread region centered around Lake Superior. In Canada the region extended from the Saskatchewan border across the southern half of Manitoba and northwestern Ontario to Lake Nipigon and the Georgian Bay region and eastwards to the Ottawa River. In the United States, it included Michigan, Wisconsin and Minnesota. Throughout the Lake Superior watershed and beyond, the Ojibway had convenient access to Betula papyrifera, usually referred to as the white, paper or canoe birch. The bark of the birch tree found extensive use in the daily life of the Ojibway and was quintessential to the traditional culture of the Algonkians of the northern woodlands.

The outer layer of birchbark, when removed in the spring, is very flexible and its inner surface, the cambium side, is smooth, soft and impressionable. It provides an ideal surface for marking with a hardwood, bone or steel stylus. Because of its ready availability, its flexibility and lightness, and its resistance to moisture, fungi and insects, birchbark was the favoured medium for Algonkian pictography. It was almost exclusively through this medium that the complex rituals and oral traditions of the southern Ojibway were stored and transmitted.

The usage of the term pictography in this paper follows that of Selwyn Dewdney (1975, p.12). It is used as a generic term for any form of preliterate art -- executed on any

available surface -- that is known, or is assumed to have
communicative rather than a decorative or aesthetic intent.
It is important to distinguish between pictography, in this
sense, and writing. The pictography to be discussed is mnemonic and does not represent the written word. Rather, by
means of its symbolism, it provides a means of recalling
corresponding oral traditions.

Dewdney (1975, pp.12-13) briefly mentions the variety of
media available and in actual use in Algonkian pictography.
It frequently appeared on the diaphragms of hide drums or
carved into the wooden sides of water drums. Sometimes song
records were inscribed on wooden slabs although birchbark was
preferred for this purpose. Ojibway communities in northern
Minnesota carved census records in wood, using totemic signs
to identify the family head and tally marks to indicate the
number in his family. The Naskapi of Labrador and eastern
Quebec stamped abstractions in color on their hide shirts
with symbolic as well as decorative effect. Some early treaties signed with Europeans were signed on parchment with
goose quill pens by native leaders using their pictographic
identity marks. George Copway (1851, pp.129-131) claimed
that the Ojibway had three depositories near Lake Superior of
"records written on slate rock, copper, lead, and on the bark
of birch trees". Although other references to records on
slate and lead have not been reported there is a description
of one on a copper plate.

William W. Warren (1957, pp.89-90) writes of a circular
copper plate containing a geneological record which belonged
to an Ojibway chief of the crane clan. He had occasion to
observe it when it was displayed to his father. On the plate
were eight deep indentations said to indicate the number of
generations of the chief's ancestors who had passed away

since they had first taken possession of the adjacent region. The crude figure of a man with a hat on its head was placed opposite the third indentation to denote the period when the Europeans made their first appearance.

The Ojibway also employed pictographs and tally records on wooden slabs used as grave markers. Henry S. Schoolcraft (1851, Part I, pp.356-357, pl.50) discusses and illustrates several examples of such grave markers. These typically contain family or clan symbol, inverted to denote death, and a variety of horizontal tally strokes which marked the frequency of important events in the life of the deceased. These events could include one or more of the following: the number of war parties led by the deceased, the number of wounds he received in battle, the number of enemies he had killed in battle, the number of peace treaties he had attended, and the number of eagle feathers he had been awarded for bravery.

Tally records are also known to have been used in time-keeping. This is well illustrated by the following passage recorded by Frances Densmore (1929, p.119). The information was provided by an Ojibway woman, Nodinens, who was 74 years old when the data was collected.

"When I was young everything was very systematic. ...My father kept count of the days on a stick. He had a stick long enough to last a year and he always began a new stick in the fall. He cut a big notch for the first day of a new moon and a small notch for each of the other days."

Despite this variety of media available for record making, the popularity of birchbark far outstripped that of all others combined. A simple example of a birchbark tally record is depicted in figure 7.1. This shows the outline of a Midéwegun, or medicine lodge, in which the Midéwewin, or

"Grand Medicine Society" as it has been labeled in English, conducted its curing ceremonies. The drawing exhibits 8 vertical strokes on its upper border and 5 on its lower border. The markings were a special notation added to a much larger pictographic scroll, illustrated by W.J. Hoffman (1891, pl.III). The notation indicated that its owner was the chief Midē shaman of his local society for 13 years.

The birchbark scroll in which figure 7.1 serves as a footnote is an example of a type of pictographic record of which several examples are extant. These were records prepared to assist in the preservation and teaching of the traditions of the Midēwewin. They form a body of pictographic material in which one can find various tallies and graphic notations exhibiting a ritual use of number. However, before exploiting this material, it is important to have some understanding of the nature of the Midēwewin and its teachings as well as some idea of the ritual use of number among the Ojibway. These items are discussed in the next few sections.

THE MIDĒWEWIN

Harold Hickerson (1970, p.52) describes the Midēwewin in the following words:

"... the Midewewin was a set of ceremonials conducted ... by an organized priesthood of men and ... women ... who had occult knowledge of "killing" and "curing" by use of herbs, missiles, medicine bundles, and other objects which had medicinal properties. Among Chippewa [Ojibway], ... members of the Mide society were repositories of tribal traditions, origins, and migrations integrated in systems of myth and legend, that is, folk-history, much of the lore being transcribed in pictographs on birchbark scrolls considered sacred. The society also owned songs and dances that would be

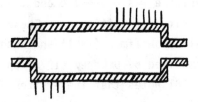

Fig. 7.1. Tally record inscribed on a birchbark scroll (after Hoffman 1891, pl.III).

performed ritually on the occasion of the meeting of the
society. A characteristic of the Mide society was limited
membership. Initiation could be achieved only after a long
period of instruction in its mysteries provided by one or
more members of high standing, often in exchange for articles
of value. Another characteristic was the ranking of members
in "degrees" ..."

The herbal knowledge of the Midéwewin could be used to
prolong life, particularly when accompanied by right living.
This medicine was considered to be a gift of God who felt
compassion for his people who were sick and dying. According
to James Red Sky Senior (Dewdney 1975, pp.23-36), God called
a meeting of the archetypal spirits of the birds and animals,
known as the Manito Council, to discuss his concern for the
suffering Ojibway. It was decided to send mankind the gift
of medicine and Bear, the most powerful of the manitos, was
chosen to bring the message to Earth. Bear completed the
first part of his journey, after four attempts, but after
reaching Earth was blocked by a large body of water. At this
point he transferred the message to Megis (Shell), or in some
versions to Otter, who carried it across the water to the

Indians on the other side.

The gift of medicine was preserved and transmitted by the Midéwewin through rites of initiation. These rites have been discussed in detail by Hoffman (1891) and Ruth Landes (1968) and have been summarized by Dewdney (1975, pp.83-84). Densmore (1929, pp.87-88) notes that these rites were intended to inject "spirit power" into the candidate which could be renewed at annual ceremonies. The spirit power was injected by "shooting" it from the medicine bags of the members. On receiving this power the candidate fell to the ground unconscious. The spirit power was conveyed by means of a small white shell (the megis) which was carried in the medicine bags of the Midé shamans.

There were four degrees of initiation which a candidate had to pass through before achieving the status of a Midé master. In some areas, four additional degrees were added, constituting the Sky Midéwewin (Landes 1968, pp.96-97). These repeated the first four grades but the locale of the ceremonies was transferred from mythic Earth to mythic Sky. The Sky ceremonies arose not from the original vision but because a patient continued ill through all Earth grades of therapy and the shamans would not abandon him. They argued that additional grades of power were available because Sky Supernaturals had supported the deliberations of the Manito Council. The patient was eventually cured and Sky rites were established, modeled upon those of Earth, but patronized by Sky creatures, Shell being replaced by Great Spirit and Bear by Eagle.

Hickerson (1963) has argued that the Midéwewin was not an aboriginal ceremonial, but developed in post-contact times. He views it as a nativistic movement which was a creative response to the stress resulting from the changing relations

between the Ojibway and the outside world brought about by
the early contact period. Nevertheless, he accepts that the
Midé cult was derived from ancient practices and beliefs
although transmuted into new ideological and ritual contexts
which would tend to include material from outside cultures.
Thus, despite a possible, or even probable late development
and the presence of some European influences, the Midé ceremonial and its ritual paraphernalia can still be regarded as
authentically Ojibwayan.

As noted earlier, the records and teachings of the Midéwewin were inscribed on birchbark scrolls. Dewdney (1975,
pp.183-191) has tabulated and labeled 137 of these and provides line drawings for most of them. He has divided the various Midé instruction charts, excluding song records, into
the following categories.

1. Origin scrolls symbolizing the origin traditions of
the Midéwewin.

2. Migration charts summarizing and charting the later
accumulations of origin lore.

3. Master scrolls and ritual charts for instruction in
the lore and rites preparatory to initiation.

4. Ghost Lodge and Sky Degree scrolls for instruction in
specialized Midé rites.

5. Deviant scrolls suspected, as unorthodox Midé scrolls,
of being used for destructive sorcery.

6. Enigmatic scrolls reflecting the results of the Midé
diffusion and decline.

Many of the Midé scrolls exhibit features having interesting numerical properties and some of these will be considered
later. The scrolls which have been selected for study will
be designated by code letters devised by Dewdney.

THE RITUAL IMPORTANCE OF THE NUMBER 4

The association of the number 4 with world directions is common throughout North America and is found among the Ojibway. Hoffman (1891, p.166) reports that according to tradition, their tribal ancestors "occupied the four quarters of the earth - the northeast, the southeast, the southwest, and the northwest". Densmore (1929, p.87) was informed that in Midé beliefs there was a Grand Medicine Spirit below which were four manitos, one at each of the cardinal points. Similarly, Landes (1968, p.194) writes that after death the "shadow" of the departed had to make a journey to the "land where midéwewin sounds forever, without end". This journey had certain perils symbolized by four evil Supernaturals associated with the cardinal points. However, the usage of 4 as a ritual number goes far beyond this central concept and can be found in such diverse areas as the games of children, puberty custons, funerary practices, origin tales, and ceremonies of the Midéwewin.

For example, in the windigo or cannibal game, a favourite of Ojibway children, a child was chosen by lot to play the part of the windigo. This was done by preparing four sticks, one longer than the others. These were held in the hand with the tops even and offered for choice among the older boys. Whoever drew the longest stick acted the part of the windigo (Densmore 1929, p.70).

Concerning puberty customs associated with young girls, Densmore (1929, pp.70-71) states that at the time of her maturity a girl was required to isolate herself for four days and nights. In addition, during the first summer of her womanhood, the girl was not allowed to take any fruit, berries or vegetables until the proper ceremony had been enacted. This began with the girl gathering the specific product of

nature for a feast to which her parents invited the Midé
shaman and others. The shaman drummed and sang, then he held
nature's bounty to her lips, but as she was about to take it
he withdrew it. This was repeated four times; on the fifth
occurrence she was allowed to eat. The same procedure was
repeated with the first of every product of nature, from the
strawberries of early summer to the wild rice of the autumn.

An interesting usage of four in connection with a love
charm is described by Densmore (1929, p.108). The charm consisted of two figurines made of wood representing a man and a
woman. These were about an inch in height and were tied
together with a hair or a raveling from the clothing of the
person to be affected. With the figurines was tied a tiny
packet of "love-charm" medicine. The figurines thus prepared
were placed in a little bag and carried by the person wishing
to create the influence. It was said the charm would attract
a person from a considerable distance, and that it could be
prepared with special herbs in such a manner that "in four
days the man to be influenced will suffer a headache so
severe as to cause nosebleed".

With respect to funerary practices, Densmore (1929, pp.74-
75) writes that the Ojibway believed everything necessary for
life and its occupations awaited a deceased person in the
"Hereafter". It was only necessary to make provision for the
deceased's comfort for the four days required to reach that
distant place. It was the custom to place food beside a
grave and to keep a fire burning there to aid the spirit, for
four nights. Landes (1968, p.191) notes that at Bois Fort,
Minnesota the Ojibway kept the dead for four days in the hope
that the soul in the spirit world would return and the person
come back to life. Elsewhere, Landes (1968, p.197) notes
that on the soul's journey after death it meets an old woman,

"Our Grandmother" who directs it further. Later the soul
encounters four old men, "Our Grandfathers".

Densmore (1929, pp.77-78) also describes the custom of
keeping a "spirit bundle" of a deceased relative. The person
wishing to do this cut a lock of hair from the back of the
dead person's head soon after death occurred. The lock of
hair was wrapped in birchbark and formed the nucleus of the
spirit bundle. The relatives built a fire on the night of
the burial, took this spirit bundle to the fire and "sat and
talked"; then they took the spirit bundle home with them.
This was repeated for four consecutive nights.

The importance of the number 4 is also revealed in origin
tales. In one such tale recorded by Hoffman (1891, p.172),
it was said that the Good Spirit first created four people
(two men and two women) who had no power of thought. They
were then made into rational beings, paired and multiplied,
and the people were placed on Earth. Between the position
occupied by the Good Spirit and Earth were four lesser spirits. When the people on Earth began to experience sickness,
misery and death, the Good Spirit and the four lesser spirits
met in the Manito Council to discuss the problem. To this
meeting were called the four wind gods. This was the gathering at which the decision was made to send the gift of medicine to the suffering Ojibway.

In another version of the origin tale, Landes (1968,
pp.90-91) was informed that the Maker sent four men to do the
work necessary for making the world. Two men worked on the
Earth and two on the Sky. In addition, the first man and
wife had four children, one of whom died - a fourth son. It
was also said that the Great Spirit made two animals and two
fowls, manitos to serve him. These manitos were responsible
for creating all subsidiary manitos associated with the

Ojibway. When the first person was created from earth by the Great Spirit, he was placed on Earth and told: "Now take four breaths". This action resulted in the establishment of air. Told to breathe four times more, he did so and established the heavens.

In a tale, telling of how the Grand Medicine was brought to the Ojibway, it was related that during the winter hunt a young male child who had accompanied his family died. The parents were much distressed and decided to return to their village and bury the body there. The family included an adopted child who was really the Sun Spirit. The adopted son informed his parents that he could bring his dead brother to life. He urged the party to hasten to the village where he had the women make a bark wigwam in the middle of which was placed the body in a covering of birchbark. The next day the family and friends were in the wigwam seated around the corpse. Eventually a bear approached, entered and placed itself before the dead body. It said hŭ, hŭ, hŭ, hŭ when it passed around the body towards the left side, with a trembling motion, and as it did so the body began quivering. The quivering increased as the bear continued until it had passed around the body four times at which point the body came to life and stood up. After this miraculous event, the bear (who seems to have been the adopted son) remained among the people and taught them the mysteries of the Grand Medicine (Hoffman 1891, pp.172-173).

A variant account of the coming of the Midēwewin is given by Densmore (1929, p.93). It is said that four manitos with the colors of the dawn painted on their foreheads came out of the eastern sky, each carrying a live otter in his hand. They used these otters in the same way as medicine bags were used in the ceremonies of the Midēwewin. By this means they

were said to have restored to life a young man who had been
dead eight days. They instructed the Ojibway to continue
this custom.

In Red Sky's version, the gift of Medicine was transmitted
to the people through the efforts of Bear and Shell. Dewdney
(1975, p.33) emphasizes that in this tale Shell "sighted land
on his fourth emergence, just as Bear broke through the
fourth barrier, and God make four attempts to create the
world".

In yet another version, it is said that a younger brother
received the revelation and used it to cure his elder brother
who was dying. The younger boy and his father made the necessary preparations, after which the boy sat down and began
the ritual. Four manitos entered in succession and shot the
dying brother [with their megis shells] who responded to the
treatment by jumping to his feet cured (Landes 1968, p.110).

In a longer account of this tale, the younger brother who
was named Cutfoot, disappeared when he was six or seven years
old and was believed dead. However, in reality, he was being
taught Earth Midéwewin by the Shell manito. He returned to
his family after four years and remained with them a year,
after which he disappeared again. One night four years later
he returned again. Afterwards, Cutfoot married and told his
wife where he had been and what he had learnt. When his
elder brother became ill and was about to die, Cutfoot put
him through Midéwewin and saved him. From then on, he taught
the old men how to perform it (Landes 1968, pp.110-111).

In the rites associated with the Midéwewin there is a conspicuous usage of the number 4 in most ritual contexts. For
example, an important part of the curing ceremony consisted
of a prior sweat bath by the senior participants. The sweat
lodge contained four stones, three of which were smaller and

flatter. These three smaller stones were first heated in a
fire outside the lodge and then placed in the middle of the
lodge. The fourth stone, as spherical as could be found, was
heated as hot as possible and was placed on the three smaller
stones. Water was sprinkled on the upper stone to create
heat. Four men usually went into a sweat lodge at a time and
the lodge was of the smallest dimensions possible for this
use (Densmore 1929, p.94). Elsewhere, Landes (1968, p.118)
notes that the construction of the lodge required four curv-
ing sticks. She also describes a larger sweat lodge formed
of eight arching sticks and containing eight stones.

As another example, one may consider the importance of the
number 4 in the activities performed by the Midé candidate or
patient. Warren (1957, p.265) writes that the person wishing
to become an initiate must choose four initiators from the
wise old men of his village. For four nights, before the
ceremony was to be performed, the medicine drums of the initiators are sounded and songs and prayers are addressed to
the master of life. Basil Johnston (1976, pp.85-93) adds
that the candidate fasted and prayed during the four days
before the ceremony. Then, on the appointed morning, the
candidate was conducted by his sponsor and tutor to the Midéwegun. At the entrance to the outer enclosure, they were met
by four bears, emblematic of the good in life. The candidate
and his sponsor, escorted by the four bears, proceeded around
the Midéwegun in four circuits before gaining admittance to
the interior. During these circuits the entourage encountered various temptations and evils which obstructed their
way. In the first degree, the evils were represented by four
other bears, in the second degree by a snake, in the third
degree by four great lynxes and in the fourth degree by a
variety of six different animal Supernaturals.

The dominance of the number 4 in the Midéwewin is manifest in the existence of the four lodges of the Earth Midéwewin (and the four lodges of the Sky Midéwewin) corresponding to the four (or eight) degrees through which a candidate passed in order to gain a full membership. The emphasis and significance of the number is apparent in Johnston's summary of the achievement of the candidate who has passed through all four degrees. "A Midewewinini [male candidate] confirmed as a fourth order member has submitted to four initiations; four times he has been purified; four times tested; four times lost his life; and four times regained it. As he left the Midewigun and the sacred posts were taken down, the medicine man or woman of the fourth order was whole and complete."

NUMBER SERIES IN THE RITES OF THE MIDÉWEWIN

An interesting feature observable in descriptions of the rites of the Midéwewin is the occurrence of numerical progressions. These result from repetitions in the mythic environment and ritual activity belonging to successive degrees of initiation into the society. Since the initiate advances through a sequence of four (sometimes eight) lodges, the number series generated consist of four (sometimes eight) terms.

One such progression derives from the fact that the Midé curing rituals required several associates working together. Landes (1968, p.114) notes that in the 1930's, "the number of colleagues varied with the locality and the midé grade, being fewer in lower grades everywhere. First grade usually required a band of about ten, second grade had sixteen, third had twenty at Cass Lake [Minnesota] (but 32 at Red Lake [Minnesota]), and so on in multiples or parts of four... The

... principal midé [shaman] ... was, during ritual activities
Shell. At first grade he had one assistant leader; for each
successive grade, he added one ... Red Lake increased the
number of assistant leaders so that at the fifth grade there
were five assistants, and so on through eighth grade. But at
Cass and Leech Lakes [Minnesota], count recommenced with the
fifth grade which was the first grade of Sky midéwewin."

From this description it can be seen that there were progressions in the number of associates and assistant leaders
required as one proceeded through the degrees of initiation.
In the case of Cass Lake the progression ran 10, 16, 20, ...
while at Red Lake it ran 10, 16, 32, ... The continuation
of these series is not listed but it is mentioned that they
continue in multiples or parts of four. The corresponding
series for the assistant leaders was more regular and ran 1,
2, 3, 4 through the four Earth degrees. At Cass and Leech
Lakes that sequence was repeated for the four Sky degrees,
but at Red Lake it continued upwards with 5, 6, 7, 8.

Densmore (1929, pp.90-92) illustrates an example of a
birchbark scroll representing four degrees of the Midéwewin
in which the number of advisors and leaders in each of the
lodges are tallied. The series of advisors runs 4, 9, 12, 21
and that of the leaders runs 1, 2, 3, 4.

In considering the number of colleagues assisting at an
initiation it is well to keep in mind Hoffman's (1891, p.168)
assertion that the officials at the second initiation are of
a "higher and more sacred class of personages than in the
first degree; the number designated having reference to quality and intensity rather than to the actual number of assistants". No doubt a similar remark is applicable to all of
the higher degrees. The first degree is apparently excepted
for it seems to have demanded four, at least as a minimum

requirement. From Hoffman's statement it is clear that the series of advisors indicate a progression in the supernatural power of the officials presiding at successive initiations. This could be achieved by increasing the number of officials participating or by employing officials of higher spiritual status.

A second type of numerical progression connected with the Midé rites arises from a central feature of the lodges corresponding to the degrees of initiation. These contained 1, 2, 3, or 4 sacred posts, or "trees of life", depending on the degree of the lodge. The sacred posts were believed to house Midé spirits whose power was invoked by the sick man (initiate). The series of sacred posts 1, 2, 3, 4 is invariant in all descriptions of the Midéwewin. Further, it is implicit in comments of Landes (1968, p.130) that in Minnesota the series of posts extended through 5, 6, 7, 8 for the appropriate degrees of Sky Midéwewin.

Other series encountered in the Midé rites derive from the Supernaturals associated with the various grades. For example, Densmore (1929, pp.90-92) describes a scroll in which evil spirits, represented by men, are stationed outside the lodges. The number of evil spirits is 4, 6, 8, or 10 according as to whether the lodge is of 1, 2, 3, or 4 degrees. In the same source, it is stated that the candidate was to bring 1, 2, 3, or 4 dogs to the initiation, the specific number depending on whether he aspired to the 1st, 2nd, 3rd, or 4th degree. The dog(s) was (were) killed and laid at the door of the lodge to which entrance was sought.

Hoffman (1891, pp.175-178) discusses another scroll in which manitos guard the various lodges against the entry of evil spirits during the night. The sequence of manitos runs 8, 12, 18, 24 increasing in agreement with the degree of the lodge.

Progressions are also associated with the custom of paying initiation fees for induction into the Midéwewin. These fees became more substantial as the degree sought increased. According to Hole-in-the-Sky, an informant of Landes (1968, p.131), "there should be at least four major items to fee first grade, eight for second, twelve for third, sixteen for fourth". This yields the arithmetic series 4, 8, 12, 16 consisting of multiples of the sacred number 4.

Landes (1968, p.135) also discusss the practice of issuing invitation sticks to the chief officers participating in the Midé ceremonies. These formal sticks were prepared by the patient during preparatory sessions. They were then issued by him to the chief officers at the head shaman's direction. Four sticks were issued at first grade, eight at second, twelve at third, and sixteen at fourth, generating the arithmetic series 4, 8, 12, 16. The introduction of invitation sticks is attributed to the following legend, related by Landes (1968, pp.135-136).

"When the midé manito commenced, he knew the Indian would not remember invitation by finger, that directly the Indian turned away, he would forget. "I guess this [stick] will be better" [thought the manito]. Therefore [Bear] reared and bit a piece off the midé Tree [cedar] so that splinters fell about. He dropped down, picked them up, counted, and there were eighty. Shell said, "Too many". Bear answered, "We will set a number for each Layer [grade]. This is what the Indian will use and henceforth cannot forget when people give sticks to him. Indeed, this will the Indian use."

The appearance of the number 80 in the legend is of some interest. Its particular selection by the story teller is perhaps accounted for by the cultural imperative to choose a relatively large number which was both a multiple of 10 -- a round number in the Ojibway decimal numeration system -- and a multiple of the sacred number 4. In addition, it may be noted that the total number of invitation sticks required for Earth Midéwewin, according to Hole-in-the-Sky's description, is the sum of the arithmetic series 4, 8, 12, 16, that is 40. If one assumes that another 40 sticks were required for Sky Midéwewin then one obtains a full count of 80 sticks in agreement with Bear's figure.

In the remainder of the paper, several example of birch-bark scrolls will be considered. These will provide graphic expression of the ritual importance of the number 4 and will also illustrate the usage of number series in the Midé cult.

POSSIBLE MIGRATION CHART RSN-I

The birchbark scroll illustrated in figure 7.2 is now in the Royal Scottish Museum in Edinburgh. It was acquired from the Denver Art Museum but is of unknown provenance.

The scroll opens with four footprints, the footprints of Bear, leading to a pair of concentric circles symbolizing the Manito Council. Four figures are stationed around the inner circle at the cardinal points and four others are similarly placed around the outer circle. Adjacent to each of the four outer figures is a tree. Four of the figures (possibly the inner four) represent the major Midé manitos mediating between God and Earth while the other four represent the wind manitos. To the right of the council circle is Bear with a bow and four arrows. He has just emerged from the Council and is about to undertake his journey to Earth. Next appear four barriers, four of Bear's footprints and four cedar trees. These illustrate Bear's path to Earth and the four breakthroughs he had to accomplish before reaching his goal.

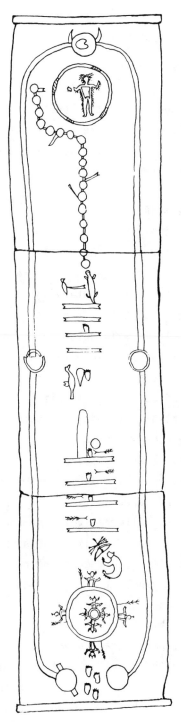

Fig. 7.2. Possible migration chart RSM-I, now in the Royal Scottish Museum (from Dewdney 1975, fig.60).

There are four other barriers to the right which probably
symbolize the four degrees of initiation in the Midéwewin.
Indeed, this is suggested by the lone Bear print after the
second barrier. It recalls a remark of Hoffman (1891, p.169)
that after the second degree and to enter the third degree
the candidate personates the bear and continues to do so
should he enter the fourth degree. Following this presumed
institution of the Midéwewin, the gift of medicine is transferred to what looks like Otter. Concerning the next stage,
Dewdney (1975, p.79) writes: "Beyond lie eighteen stopping-
places, where the Midé message is delivered, alternatively
interpretable as a migration route with misleading offshoots,
or simply as the Path of Life". This sequence of stopping
places leads to a circle with hatching in four sections. The
circle contains a figure marked with short lines undulating
from the head and others radiating from the body. This
characterization indicates high spiritual power and seems to
represent the potential attainable through the Midéwewin.

The activities recorded in the scroll occur within a formalized border which Dewdney (1975, pp.73-75, 79) suggests
may be the Great Midéwegun, a term he uses to denote Lake
Superior and the lands around it, briefly the territorial
world of the Ojibway. Alternatively, or concurrently, it may
represent the external path of life on which men must travel
which can be contrasted with the interior life symbolized by
the Midéwewin. The symbol at the far right in which the
border terminates has been described as "End of the Road".
Red Sky has termed this "Everlasting Life". It exhibits two
horns symbolic of supernatural power.

The notations in the scroll show the considerable ritual
significance attached to the number 4. In this instance it
has been graphically emphasized in no less than ten ways.

MASTER SCROLL KP-1

The scroll depicted in figure 7.3 comes from Milles Lacs, Minnesota and was collected in 1964. It can be recognized as a master scroll by the presence of four rectangular floor plans corresponding to the four lodges of the Midéwewin. There are 4 officials shown in the first lodge, 8 in the second, 16 in the third, and 36 in the fourth, yielding the sequence 4, 8, 16, 36. The inner rectangle in each lodge represents the path of the procession around the interior of the Midéwegun, made at intervals during the ceremonies. Within each lodge there are four manitos, probably bears, two guarding each entrance. Lurking between the lodges are evil manitos which block the entrances. The space between the first and second lodges is dominated by the Great Snake, that between the second and third by the horned Misshipeshu, the Great Lynx, and that between the third and fourth by a great two headed horned manito. In addition, the first three lodges are each blocked by four manitos, two near each entrance. The fourth lodge is uniquely surrounded by twelve bird-like figures, possibly Sky manitos, with an additional anthropomorphic horned figure and two bear figures. Somehow, these seem to represent benign manitos although verification is lacking.

The network of lodges and manito figures is encompassed by a border which seems to represent the Great Midéwegun, the Ojibway universe. It begins on the left with a small circle containing Bear and terminates on the right with the horned symbol of Everlasting Life. Along its upper border, close to the upper left hand corner of each lodge, is a representation of Bear seated before his sacred drum. This emphasizes the prime importance of Bear as patron manito of the Midéwewin. Outside of this border, and supporting the notion that it

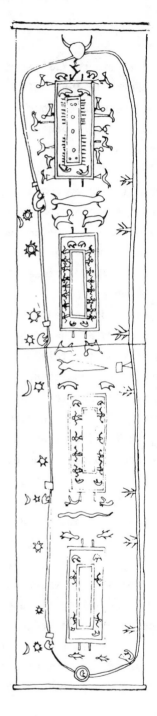

Fig. 7.3. Master scroll kp-1, now in the private collection of Karen Peterson (from Dewdney 1975, fig.81).

marks the Ojibway world, are lunar and solar symbols. Along
the lower border is a sequence of trees which represents a
forest. The triangle with rectangle surmounting it on the
lower border below Misshipeshu seems to be an oblique reference to a bad Midé shaman, "one who employs his power for
evil purpose" (Dewdney 1975, pp.131-133).

It may be noted that this scroll also emphasizes the
sacred number 4 in many ways. This can be seen in the number
of lodges, the number of bears in each lodge, the number of
evil manitos adjacent to the entrances of the first three
lodges, the number of bird manitos about the fourth lodge
(12 = 4 × 3), the number of Bear and drum configurations on
the boundary, and the number of lunar symbols. Moreover, the
single number series encountered is based on multiples of 4.

MASTER SCROLL DE-1

A highly formalized birchbark scroll is shown in figure
7.4. The original was reproduced and interpreted by Densmore
(1929, pp.90-92, pl.34). The scroll is in an unusual geometric style but can be identified as a master scroll by the
representations of the four lodges of the Midéwewin. Following Densmore's comments, the sequence of tally marks along
the lower half of the processional path represents Midé officials. The sequence runs 4, 9, 12, 21. Immediately above
the upper half of the processional path is a sequence of
dots, corresponding to the principal shaman and his assistants, which runs 1, 2, 3, 4. In the central portion is
another sequence of dots which runs 2, 3, 4, 5. These are
symbolic of a sacred fire and 1, 2, 3 or 4 sacred posts.
Along the outer edges of the lodges are postlike extensions
which generate the series 4, 6, 8, 10. These are reputed to
be evil spirits who try to influence the candidate and during

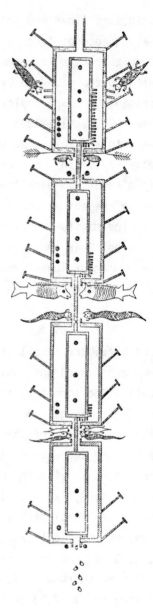

Fig. 7.4. Master scroll De-1 (from Densmore 1929, fig.12).

the ceremonies are represented by men. In addition, various
evil manitos are shown lurking between the successive lodges.
With the exception of the series of Midé officials, the number series in this scroll are all regular arithmetic series.

MASTER SCROLL MHSM-2
     The scroll illustrated in figure 7.5 is now in the Minne-
sota Historical Society Museum in St. Paul. Its provenance
and date of collection is unknown. The interpretation given
here is due to the author and is based on its similarity to
other scrolls described in the literature. It is apparent
from the four lodges that this is a master scroll. The
fourth lodge is different from the first three in that it
consists of crossed lodges. This is probably a graphic
device to illustrate that the lodge of the fourth degree has
four entrances, a common tradition despite its absence in the
two master scrolls already considered. The activities taking
place begin on the left with a stylized symbol of the Manito
Council with four square projections marking the cardinal
points. To the right are four footprints of Bear signifying
his role in bringing the Midéwewin to Earth. Four evil manitos block the east-west doors of each of the four lodges, two
located at each entrance. In addition, the Great Snake lies
across the path between the first and second lodges, and the
great two headed horned beast lies across the path between
the second and third lodges. Eight manitos, including copies
of Misshipeshu, all but one with horns of power and all with
power emanations, appear to have penetrated the walls of the
fourth Midéwegun. Two are shown at each of the entrances,
where they hover menacingly, partially within the lodge.
Perhaps to combat this threatening evil, four bear manitos
are stationed within the lodge, one in each quarter. A

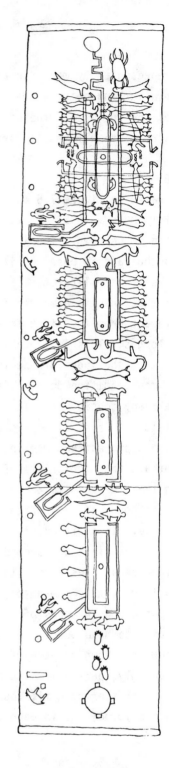

Fig. 7.5. Master scroll MHSM-2, now in the Minnesota Historical Society Museum (from Dewdney 1975, fig.77).

smaller lodge is connected diagonally to the upper left corner of each Midéwegun and a Midé shaman with a pipe and drum
in hand is shown adjacent to each of these. It seems likely
that these smaller structures are sweat lodges in which the
principal officials were purified before entering the Midéwegun. Along the upper edge of the scroll are three depictions
of Bear with his sacred drum, illustrating Bear's role as
patron manito of the Midéwewin. At the far right is the circle of Everlasting Life and below it to the left is an enigmatic crab-like figure, possibly of evil portent.

To complete the description of the scroll it remains to
consider the sequence of Midé officials.

This runs 4, 8, 16, 16. It is interesting to note that
the 16 officials in the third lodge are given a single horn
of power whereas 15 of the officials in the fourth lodge are
given a full pair of horns. It would appear that the sixteenth, who is shown bald, was overlooked. Since horns are
symbols of supernatural power, the officials in the fourth
lodge have a power status double that of those in the third
lodge. Thus the number of officials in the fourth lodge, or
rather the power which they represent, is effectively doubled
and the sequence is equivalent to 4, 8, 16, 32. The series
in figure 7.5 is therefore a geometric series in which the
doubling in the second and third lodges occurs in the principal tally of officials and the doubling in the fourth lodge
occurs in a subsidiary tally of horns.

MASTER SCROLL MMM-1

The scroll shown in figure 7.6 was found on the west side
of Lake Winnipeg in an abandoned Midé cache at Jackhead in
1969 (Dewdney 1975, pp.102,185). It is now located in the
Manitoba Museum of Man and Nature. The scroll again shows

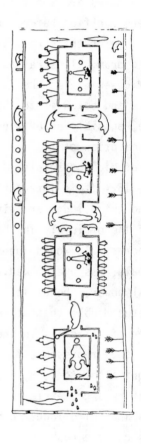

Fig. 7.6. Master scroll MMM-1, now in the Manitoba Museum of Man and Nature (from Dewdney 1975, fig. 95).

the four lodges of the Midéwewin, ordered from right to left.
Between the lodges and lurking near the entrances are manito
figures which from precedents already considered may be
regarded as evil influences. Along the upper border are
three references to the patron manito Bear and his sacred
drum, together with inscribed circles in sets of two and
four. Along the lower border is a fourth reference to Bear
with his drum and a collection of trees symbolizing the forest. Looking at the lodges, 4 officials holding rattles are
associated with the first lodge. The number of officials
rises to 8 in the second, to 16 in the third and drops to 4
in the fourth. However, each of the officials in the fourth
lodge is provided with a horn of power and are therefore of
higher status. Thus, even though the sequence runs 4, 8, 16,
4 the perception of the sequence, given the power convention,
is probably once again 4, 8, 16, 32.

This scroll deviates from the others considered in that an
anthropomorphic figure is inscribed in each of the lodges.
My reading of this scroll is that the figure represents the
candidate himself who is growing in power as he progresses
through the degrees. The power relationship is exhibited by
the characteristic treatment of the hair. In the fourth
lodge the candidate is utterly transformed becoming manito-
like. It is noteworthy that in the lower half of the fourth
lodge are four bear prints while another set of six are shown
leading out of the final exit. This suggests that the candidate leaves the fourth degree ceremony with the powers of the
bear. No longer is the emphasis on reaching the final goal,
Red Sky's "Everlasting Life". Rather, it appears to focus on
the glorification of the candidate and the acquisition of the
powers of Bear.

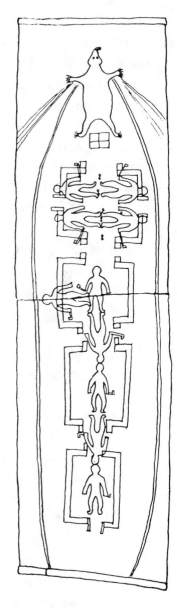

Fig. 7.7. Master scroll HF/A-5, now in the Heye Annex of the Heye Foundation (from Dewdney 1975, fig.143).

MASTER SCROLL HF/A-5

The final scroll to be considered is illustrated in figure 7.7. It was collected at Berens River on the east shore of Lake Winnipeg in 1932. It is now the property of the Heye Foundation in New York and is located in the Heye Annex. The scroll contains the customary four lodges with east-west entrances. The fourth lodge has additional north-south entrances. The presiding shamans, shown with sacred pipe in hand, occur in the sequence 1, 2, 3, 4. Here the similarity with the previously considered scrolls ends. Bear totally dominates the scene and it seems apparent, as Dewdney (1975, p.140) suggests, that the shamans "are conducting rites that lead towards acquisition of the enormous powers of the Bear himself".

The scroll in figure 7.6 appears to mark an intermediate stage in a transition from the classical scrolls in figures 7.3-7.5 to the deviant scroll in figure 7.7.

ACKNOWLEDGEMENT

This work has been supported by a research grant from the Social Sciences and Humanities Research Council of Canada (410-79-0448).

8. A Survey of Aztec Numbers and Their Uses

Stanley E. Payne and Michael P. Closs

INTRODUCTION

Late in the tenth century the legendary ruler Quetzalcoátl established Tula, in the state of Hidalgo, Mexico, as the capital of the Toltecs. The Toltecs were a Mesoamerican people whose language, related to the Ute language of the southwestern United States, became a contributor to the Náhuatl language still spoken today by the Aztecs of Central Mexico. The tribal records of the Aztecs indicate that they began their wanderings in A.D. 1168 (Vaillant 1950, p.97). After leaving their legendary ancestral home of Aztlán and wandering for many years, they eventually reached the Central Mexican Valley. In 1325, according to tradition, they founded their capital, Tenochtitlán, within the bounds of present day Mexico City (Caso 1958, p.xiv). By the fifteenth century it had become the center of Aztec growth, conquest, and expansion. When Cortés arrived in 1519, Tenochtitlán dominated all other cities and had reached the height of its power and magnificence. Cortés also found a civilization whose religious and bureaucratic needs required a fairly extensive use of moderately large numbers and their symbolic representations. It is the purpose of this paper to give a brief description of the Náhuatl number sequence, to illustrate some of the uses of numbers and numerals in the Aztec culture by presenting a few specific examples, and to discuss briefly what generally is known or has been accepted about these matters.

Although we lack details on Aztec arithmetical procedures,

we do know their number words and do have a partial understanding of their number symbols. Since their number symbols
were used primarily for calendrical and economic bookkeeping,
any discussion of numbers and their use must be accompanied
by a discussion of their calendar system and tribute records.

NÁHUTAL NUMBER WORDS

In our decimal number system, numbers are written in
positional notation using a base of 10. A symbol of the form
$b_n b_{n-1} \cdots b_1 b_0$, where $0 \leqslant b_i \leqslant 9$ for $i = 0, 1, \ldots, n$, refers
to the number

$$b_0 + b_1 \times 10 + b_2 \times 10^2 + \ldots + b_n \times 10^n.$$

For example, $276 = 6 + 7 \times 10 + 2 \times 10^2$. The same type of notation can be used with bases other than 10. For instance, in
the case of a base of 20, the symbol $b_n b_{n-1} \cdots b_1 b_0$ where
$0 \leqslant b_i \leqslant 19$ for $i = 0, 1, \ldots, n$, refers to the number

$$b_0 + b_1 \times 20 + b_2 \times 20^2 + \ldots + b_n \times 20^n.$$

This is essentially the system the Aztecs used to represent
numbers in their speech. In the linguistic formation of
numbers 1 through 19 a secondary base of 5 was used. Their
number words, introduced below, illustrate the principle.

Because of phonetic variations among the Aztecs and in the
orthography used to render Náhuatl sounds in the Spanish
alphabet, the number words appear with slight morphophonemic
variations. Moreover, when counting objects from certain
specific classes, there are corresponding modifications in
the number words used. For the linguistic data, we follow
Thelma D. Sullivan (1976, pp.189-195).

1   ce

2   ome

3   ei, yei

4   nahui
5   macuilli

This completes the first basic group, with __macuilli__ no doubt deriving from __maitl__, 'hand'. According to Manuel Orozco y Berra (1960, pp.443-444), the term may be broken down further into __cui__, 'to take', and __pilli__, 'fingers', and means something like "fingers taken with the hand".

6	chicuace	five plus one
7	chicome	five plus two
8	chicuei	five plus three
9	chiconahui	five plus four
10	matlactli	

__Matlactli__ may come from __maitl__, 'hand', and __tlactli__, 'torso'.

11	matlactli once	ten plus one
12	matlactli omome	ten plus two
13	matlactli omei	ten plus three
14	matlactli onnahui	ten plus four
15	caxtolli	

__Caxtolli__ seems to be a new basic term for which there is no known etymology.

16	caxtolli once	fifteen plus one
17	caxtolli omome	fifteen plus two
18	caxtolli omei	fifteen plus three
19	caxtolli onnahui	fifteen plus four
20	cempoalli	one counted group

"Twenty" is more than just the base of the Aztec number system. It is the "I" (the individual composed of four parts -- the feet and the hands -- each with five appendages). To form larger numbers, the first nineteen are placed before a numerical root to indicate a count of the base unit and are placed after the root to indicate addition to the base unit.

30	cempoalli ommatlactli	one score plus ten
37	cempoalli oncaxtolli omome	one score plus seventeen
40	ompoalli	two score
60	eipoalli	three score
100	macuilpoalli	five score
399	caxtolli onnauhpoalli ipan caxtolli onnahui	nineteen score plus nineteen
400	tzontli	

Tzontli means 'hair' or 'growth of garden herbs' and, in any case, signifies multitude or abundance. The word ipan, a preposition meaning 'on top of' or 'plus' is used to connect multiples of distinct powers of twenty for larger numbers, whereas the usual ligature for numbers within one multiple of twenty is on or om.

401	centzontli once	one '400' plus one
405	centzontli onmacuilli	one '400' plus five
500	centzontli ipan macuilpoalli	one '400' plus five score
7999	caxtolli onnauhtzontli ipan caxtolli onnauhpoalli ipan caxtolli onnahui	nineteen '400's plus nineteen score plus nineteen
8000	cenxiquipilli	one '8000'

Xiquipilli refers to a 'bag of cacao beans' and represents the third power of twenty. We know of no special word
name for 160,000, the fourth power of twenty. However,
numeration between 8,000 and 160,000 continues by combining
xiquipilli with the other smaller numbers. The regularity of
the system can be appreciated by considering the following
example.

$$113,197 = 14 \times 8000 + 2 \times 400 + 19 \times 20 + 17$$
$$= \text{matlactli onnauhxiquipilli ipan ometzontli}$$
$$\text{ipan caxtolli onnauhpoalli on caxtolli omei}$$

There are four nouns which are combined with numerals to
count things of various forms or types. Tetl, 'stone', is
used in counting round things and generates numerals such as
yetetl, '3', and macuilpoaltetl, '100'. Pantli, 'banner,
flag', is used in counting rows of people or things and leads
to numeral forms such as cempantli, '1', and nappantli, '4'.
Tlamantli, 'thing', is used in counting pairs or groups of
things or different kinds of things and yields numeral forms
such as centlamantli, '1', etlamantli, '3', and macuil-
tlamantli, '5'. Olotl, 'corn cob without kernals', is used
to count things which roll or turn and generates numerals
such as omolotl, '2', and caxtololotl, '15'. According to
Sullivan (1976) the only variant among the numerals is
tlamic, '20', which derives from tlami, 'to complete', and
belongs in the sequence of numerals ending in olotl.
    Tecpantli is used when counting persons or things 'by
twenties', and in such a context generates the numerals
centecpantli, '20', ontecpantli, '40', etecpantli, '60', and
so on. Ipilli is used when counting 'flat things' by
twenties. Thus, for example, in counting mats, one has

cemipilli, '20 (mats)', omipilli, '40 (mats)', eipilli, '60
(mats)', and so on. Quimilli is used for counting 'things
which enwrap' by twenties. For example, in counting sets of
twenty blankets one uses the numerals cenquimilli, '20
(blankets)', onquimilli, '40 (blankets)', yequimilli, '60
(blankets)', and so on.

Náhuatl also contains other numeral forms, such as ordinals, and includes particles and suffixes which can be used
to modify numerals to indicate a variety of specific notions
such as occur in expressions like 'two more' or 'five times'
(Sullivan 1976).

NUMBER AND CULTURE

As will happen in any society, the number system employed
by the Aztec had an impact on their mode of cultural expression. Because the primary base of the Náhuatl number system
is 20, it is not surprising to find that 20, 400 and 8000 had
special significance. We will consider a few examples which
illustrate the cultural impact which the base 20 had on the
Aztec mind. We will first look at some instances where that
influence can be seen in Aztec mythology and prehistory.

An appropriate place to begin is the legend of the birth
of the Aztec tribal god Huitzilopochtli who was identified
with the sun. It is related (Caso 1958, pp.12-13) that the
earth goddess Coatlicue, after having given birth to the moon
and the stars, retired to a life of retreat and chastity.
One day while sweeping she conceived miraculously. When her
children, the moon, Coyolxauhqui, and the stars, called Cen-
tzonhuitznáhuac, discovered that she was pregnant they became
furious and determined to kill her. Coatlicue wept over her
approaching death but was consoled by the unborn son in her
womb who spoke to her saying that when the time came he would

defend her. Just as her enemies came to slay the mother, Huitzilopochtli was born. He cut off Coyolxauhqui's head and put the Centzonhuitznähuac to flight. His victory, repeated every day at sunrise, symbolized a new day of life for men. Of interest to us is the fact that the word for the stars, Centzonhuitznähuac, signifies '400 huitznähuac' and may be interpreted as an allusion to the "multitude" of stars.

A similar usage of 400 appears in Aztec prehistory in connection with an early episode involving the 400 Chichimeca, also called Mimixcoa. To the Aztecs, "Toltec" implied everything that was oldest and most refined in Mesoamerica while "Chichimec" represented the new and the barbarous. In the Anales de Cuahtitlan (Davies 1977, p.430) it is told that the 400 Chichimeca fell under the power of Itzpapalotl (a goddess of the Chichimec) who proceeded to eat them. Afterwards, Itzpapalotl is shot with arrows, killed and burnt by Mixcoatl (a Chichimec god). In the Leyenda de los Soles (Davies 1977, p.432) it is Mixcoatl, his three brothers, and one sister who are credited with slaying the 400 Mimixcoa.

Henry B. Nicholson (1971, p.402) describes a variant account of the above events in which the Aztec god Tezcatlipoca, who is partially merged with Huitzilopochtli, created 400 men and five women. Nicholson suggests that these 400 men were the Centzon Huitznahua which he translates as the '400 Southerners'. An internecine struggle ensued and after three years of fighting the 400 men perished to provide food for the future sun, while the five women (one of whom was Coatlicue, the mother of Huitzilopochtli) perished later on the day the sun was created. In the year after the creation of the sun, the deity Mixcoatl-Camaxtli created four men and a woman to stir up fresh discord, and in the next year he struck with a staff a rocky cliff, from which issued forth 400 Chichimeca. Later, he sent the five individuals he had previously created to attack the Chichimeca-Mimixcoa, who were idling away their time in drunken revels. There then followed a mass slaughter, in which all but three (including Mixcoatl transformed) were slain.

The number 400 also shows up in deity names and is incorporated into the title of the Mexican gods of pulque, an intoxicating beverage made from the juice of the maguey plant . Munro S. Edmonson (1971, p.43, n.1166) notes that these deities were called the 400 Rabbits and quotes (in translation) the following passage from Sahagun:

> "But they said
> That wine is called 400 Rabbits
> Because there were many
> And varied ways of drunkenness."

In the same note, Edmonson also refers to a myth about 400 Cloud Serpents who were closed in a cave for four days. It is said that they were given maguey to suck and invented pulque.

The maguey itself was directly personified as a goddess, Mayahuel. Her exuberant fertility was dramatized by her representation as a female with 400 breasts (Nicholson 1971, p.420).

A striking use of the number 400 appears in a metaphor collected by Sahagun which Sullivan (1963, p.107) suggests is based on the sense of 400 as an incalculable number. She presents the metaphor in the following words.

"This is said about someone who knows a great many things, such as painting on paper, or such crafts as forging metal, carpentry, and goldsmithery. He knows all these things well. For this reason it is said: He succeeded in achieving four hundred."

It is natural to expect that the vigesimal nature of the Aztec number system would also be manifest in economic transactions. In fact, it will be seen in a later section that tributary items were customarily demanded in multiples of 20, 400 or 8000. For the moment, we may note that in view of the derivation of the term for 8000 it comes as no surprise that large numbers were required in counting cacao beans. Cyrus Thomas (1900, p.920) makes the following quotation from a translation of Clavigero's history of Mexico: "They counted the cacao by xiquipilli ... and to save the trouble of counting them when the merchandise was of great value [probably quantity] they reckoned them by sacks, every sack having been reckoned to contain 3 xiquipilli, or 24,000 nuts." The proposed change of "value" to "quantity" is not really necessary here since cacao beans were used as currency in Mesoamerica and so, in this case, the two concepts coincide.

Finally, it may be noted that the influence of the vigesimal base of Aztec numeration extends to the social organization of the Aztec community. Indeed, at the lowest levels of government, families were grouped into collectivities known as calpulli (singular, calputin). The importance of these groupings is apparent by their continued existence in the post-conquest period. Charles Gibson (1964, pp.152, 182, 514, n.103) writes: "Baptismal and other documents consistently recorded the calputin in identifying the origins and affiliations of the members of the community". Of special interest here, is his observation that: "Within the calputin, families appear to have been grouped in vigesimal units, each with its officer. Many notices comment especially on the centecpanpixqui (20-pixqui) with jurisdiction over twenty families, and the macuiltecpanpixqui (100-pixqui) or

centurion, with jurisdiction over 100 families." Later, Gibson adds that surviving records outside the Valley of Mexico
suggest that the units of 20 and 100 actually contained only
rough approximations of these numbers.

CALENDRICAL RECORDS

The Aztecs, along with other Mesoamerican cultures, used
two major calendrical cycles, a sacred almanac called the
tonalpohualli and an annual calendar that approximated the
"tropical" year. The tonalpohualli was constructed from a
sequence of twenty day names paired with the sequence of
numbers from 1 to 13. The names and hieroglyphs of the day
names are represented in Table 8.1. For successive dates the
sequence of numerical coefficients continues in increasing
order to 13 and then begins again at 1 while the day names
continue in the order listed in Table 8.1 to the last day
Flower and then begin again with Crocodile. The numerical
coefficient in each case is represented by an appropriate
number of small circles. Since 13 and 20 are relatively
prime (that is, their highest common factor is 1), their
least common multiple is 260 and so the succession of these
dates yields a calendar of 260 days.

The solar year consisted of eighteen named months with
twenty numbered days in each, followed by a short five day
period to make up a 365 day cycle. The 260-day and 365-day
calendars were combined so that each day could be specified
by both a sacred date and an annual date. Since the least
common multiple of 260 and 365 is 18,980 (= 52×365), the
combined cycle of the two calendars would only repeat after
52 years of 365 days. This 52-year cycle was known among the
Aztecs as the xiuhmolpilli, 'sacred bundle', and played a
significant role in their religious life. In fact, stone

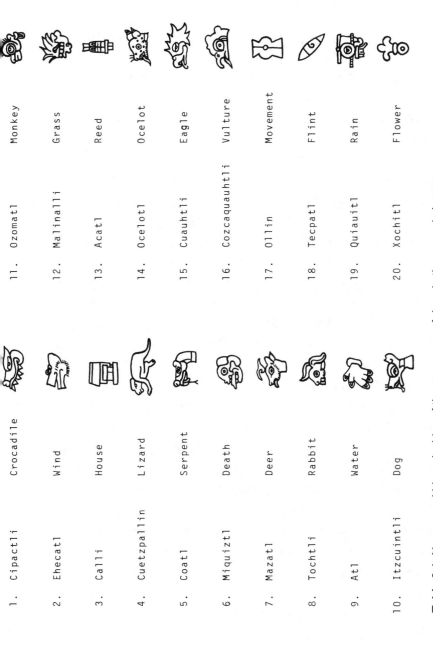

1. Cipactli	Crocadile		11. Ozomatl	Monkey
2. Ehecatl	Wind		12. Malinalli	Grass
3. Calli	House		13. Acatl	Reed
4. Cuetzpallin	Lizard		14. Ocelotl	Ocelot
5. Coatl	Serpent		15. Cuauhtli	Eagle
6. Miquiztli	Death		16. Cozcaquauhtli	Vulture
7. Mazatl	Deer		17. Ollin	Movement
8. Tochtli	Rabbit		18. Tecpatl	Flint
9. Atl	Water		19. Quiauitl	Rain
10. Itzcuintli	Dog		20. Xochitl	Flower

Table 8.1. Names and hieroglyphics of the sequence of days in the sacred almanac.

models of the xiuhmolpilli were deposited in ritual "tombs"
at the expiration of a 52-year cycle.

A given 365-day year was designated by the sacred almanac
name of its 360th day (Broda de Casas 1969, pp.35-36). For
example, the Aztec prophecies stated that the man-god
Quetzalcoátl would return in a year 1 Reed and it was in precisely such a year that Cortés first arrived in Tenochtitlán
causing the current Aztec ruler Moctezuma considerable con-
sternation. Counting through 365 days, the 360th day of the
new year would be 2 Flint. The following year would be named
3 House, the one after that 4 Rabbit, and the next one 5
Reed. It is easy to see that in the 52-year cycle only four
day names -- Reed, Flint, House, Rabbit -- actually appear as
year names. This results from the fact that 5 is the greatest common divisor of 20 and 365 so that of the 20 day names
only 20 ÷ 5 = 4 could serve as 'year bearer', i.e., as name
day for the year.

An interesting succession of years from the Mendocino
Codex is depicted in figure 8.1. A vertical column of year
names is found on the left side of the folio page starting
with 1 Flint, 2 House, 3 Rabbit, 4 Reed and running down to
13 Reed. The years express the duration of the reign of the
seated figure, Itzcoatin, fourth ruler of Tenochtitlán, which
lasted from 1 Reed (A.D. 1427) to 13 Reed (A.D. 1440). The
symbol in front of Itzcoatin consists of three spears and a
shield bearing seven eagle-down feathers, emblematic of the
Aztec capital, and denotes the authority of the Aztec lords.
The surrounding illustrations represent collapsing and burning temples together with associated hieroglyphs to which
later Spanish glosses have been added. These scenes represent towns defeated by Itzcoatin while the hieroglyphs and
their glosses indicate the name of the town involved. As

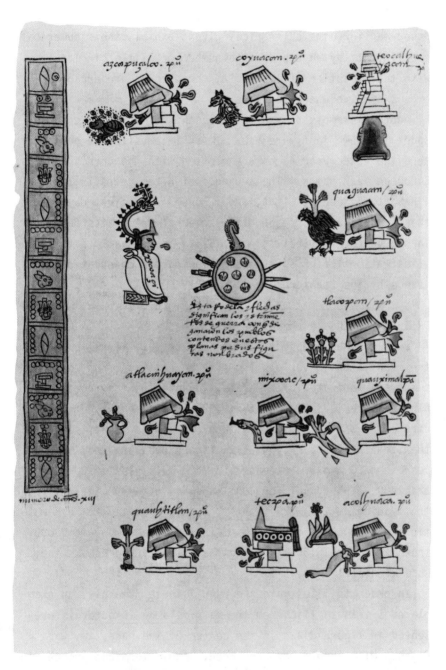

Fig. 8.1. Mendocino Codex, Plate 5 (from Bernal and León-Portilla 1974, p.296).

noted earlier, the calendrical glyphs exhibit their numerical coefficients by an appropriate number of little circles. With the exception of 11 Rabbit, the circles are arranged in groups of five where feasible.

Another example of year glyphs, this time from the Aubin Codex is shown in figure 8.2. It is interesting to note that here the Aztec year names are glossed with the corresponding European year names. The sequence of Aztec years is 3 House, 4 Rabbit, 5 Reed, 6 Flint, and 7 House. The first days of these years have corresponding sacred almanac dates falling in the years A.D. 1521 to A.D. 1525. The Náhuatl text and illustrations refer to events following the conquest of the Aztecs by the Spaniards in A.D. 1521.

TRIBUTE RECORDS

Some of the Aztec codices contain tribute lists in which the quantities of various items to be received from conquered towns are indicated by numerals. Quantities from one to nineteen are indicated by the requisite number of dots or circles as shown in figure 8.3a or by simple repetition of the item involved. Flags as in figure 8.3b are used to represent 20 and are repeated to form multiples of 20 less than 400. The symbol for 400, shown in figure 8.3c, reflects the meaning of the word tzontli, and is repeated to form multiples of 400 less than 8000. Finally, the symbol for 8000, shown in figure 8.3d, is a bag that clearly represents the word xiquipilli.

In order to illustrate the use of these numerals, an example of a tribute list from the Matrícula de Tributos is presented in figure 8.4. In the center of the page, two war dresses with shields are recorded. Above them and proceeding in a clockwise order from the upper left hand corner, one

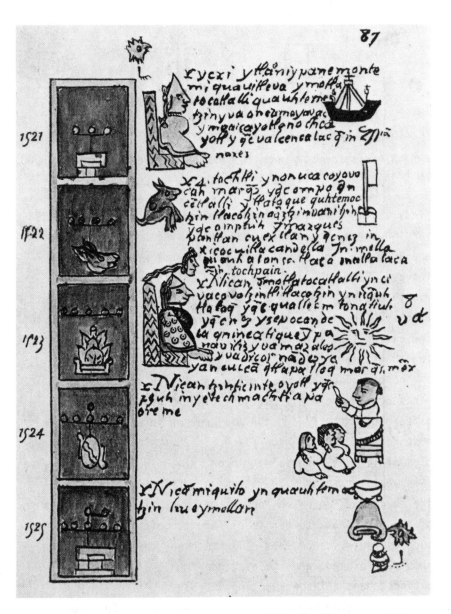

Fig. 8.2. Aubin Codex, Plate 87 (from Bernal and León-Portilla 1974, p.304).

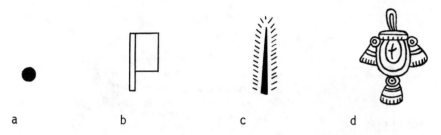

Fig. 8.3. Numeral symbols employed in the tribute records: (a) 1, (b) 20, (c) 400, (d) 8000.

finds a bin of various grains, 200 small pots of honey, 1200 varnished gourd bowls, 400 baskets of white copal incense, and 8000 balls of unrefined copal incense wrapped in palm leaves. Below the war dresses and shields, one finds 400 richly decorated mantles, 400 women's blouses and 1200 plain mantles. The latter group of mantles have two fingers on top indicating that they have a length of two brazas, twice the length of the usual mantle.

The braza appears most often as the equivalent of two varas or 1.68 meters. However, other brazas are recorded in different contexts, and the measurement intended is at times obscure. In fact, Gibson (1964, pp.257-258) points out examples of variant "brazas" of three, six, nine, ten, and twelve pies [each of about 0.28 meters] at different localities. The source of confusion appears to have arisen from the use of the term braza to denote different empirical measurements such as the distance from elbow to hand, the distance from shoulder to hand, and the distance from the foot to the extended hand (Gibson 1964, pp.538-539, n.4).

At the bottom and lower right hand side of the tribute list are hieroglyphs which are believed to represent the locations of the imperial tribute collectors (Gibson 1964, pp.34, 194-195).

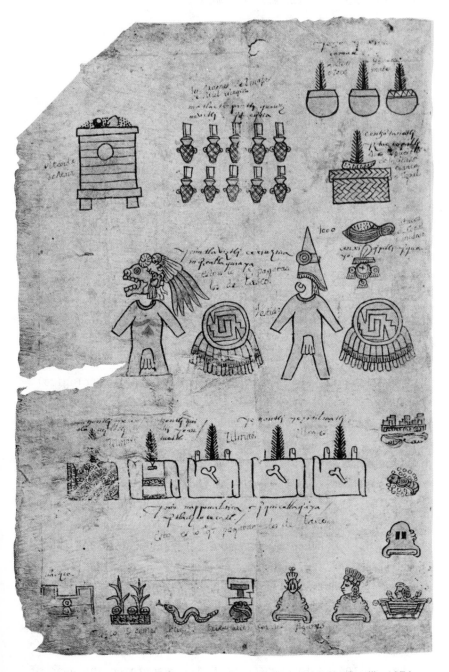

Fig. 8.4. Matricula de Tributos, Plate 16 (from Bernal and León-Portilla 1974, p.263).

Fig. 8.5. Detail from the Mendoza Codex, Plate 62 (after Ross 1978, p.89).

CHRONOLOGICAL RECORDS

The Aztecs also used numerals to represent chronological counts which might vary from a few days to thousands of years. Figure 8.5 shows a scene in the Mendocino Codex where a novice priest is being punished for negligence and insubordination by being pricked with maguey spikes. The text next to the small house glyph reads: "If the novice alfaqui went home to his little house to sleep for three days, they punished him in the manner described above" (Ross 1978, p.89). The face in the house glyph represents the novice and the three black dots in front of the face represent the three days spent at home. From the representation of three days by three black dots it may be inferred that a time count of one to nineteen days would be represented by the requisite number of black dots.

The portion of the Mendocino Codex describing the ages of youth, male and female, represents those ages by a tally of

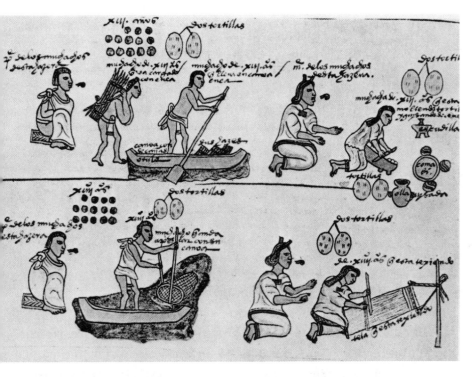

Fig. 8.6. Detail from the Mendoza Codex, Plate 60 (from León-Portilla 1974, p.290).

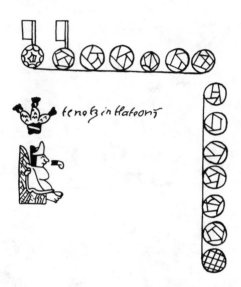

Fig. 8.7. Aubin Codex, Plate 139 (after Dibble 1963, p.135).

blue disks, each blue disk representing one year. The number
of blue disks rises from 3 to 15 in accordance with the age
of the youths. For example, figure 8.6 illustrates what is
required of 14 year old children. Below 14 blue disks representing the chronological age of the children, the father is
shown in charge of his son who is learning to fish while the
mother is shown in charge of her daughter who is learning to
weave. Above both the son and daughter are two tortillas
indicating the daily ration of maize at this age in life.

There is reason to believe that the color usage is significant here. In fact, the year glyphs discussed earlier are
also painted blue and it suggests that this color was preferred for chronological references to periods of 365 days.

A larger chronological interval of 52 years appears in the
Aubin Codex and is reproduced in figure 8.7. Here, the two
disks with flags each symbolize 20 years while the remaining
12 disks each symbolize one year.

Fig. 8.8. Detail from the Mendoza Codex, Plate 71 (after Ross 1978, p.109).

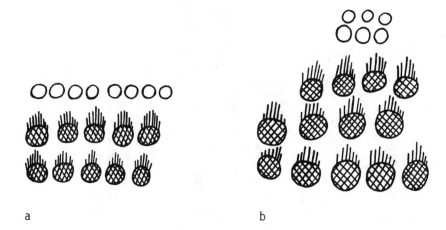

a   b

Fig. 8.9. (a) Detail from the Vatican Codex, Plate 7 (after Thomas 1900, fig. 38).
(b) Detail from the Vatican Codex, Plate 10 (after Thomas 1900, fig. 39).

The largest chronological interval of this type, namely 70
years, occurs in the Mendocino Codex and is shown in figure
8.8. The manuscript at this point deals with the theme of
drunkenness. The illustration gives the chronological notation for 70 years, literally the proverbial three score years
and ten. The central figures in the accompanying scene are
an elderly male and female who have reached this age and who
are now allowed to drink the fermented juice of the maguey
plant, a privilege denied to those who are younger.

The Vatican Codex contains other chronological counts
which measure much larger intervals. Thomas (1900, p.947)
notes that this codex contains the symbols in figure 8.9a,
which are interpreted as 4008 years and which refer to the
years of the second age of the world. Each one of the cross-
hatched and fringed disks, blue in the original, represents
400 years while the upper row of 8 smaller disks, also blue,
represents 8 years. The cross-hatched disks may be compared

with the disks in figures 8.6-8.8, also crossed in various manners and colored blue. The fact that the present disks are also fringed suggests that they are a conflation of the symbol for **tzontli**, 400, and the blue disk representing a period of 365-days. Thomas also considers one other example of this type shown in figure 8.9b. In this case, the symbol is interpreted as a reference to 5042 years. However, as Thomas points out, based on the above interpretation, this should be corrected to (13 × 400) + 6 = 5206 years.

LAND RECORDS

Harvey and Williams (1980) have shown that a system of numerical notation, distinct from those described earlier, was employed in land documents of the Texcocan-Aztecs. Their results are discussed in detail in the following paper and will not be dealt with here.

ACKNOWLEDGEMENTS

We thank Herbert R. Harvey for his helpful comments on the original version of this manuscript.

The participation of Closs in this work has been supported by a research grant (410-79-0448) from the Social Sciences and Humanities Research Council of Canada.

9. Decipherment and Some Implications of Aztec Numerical Glyphs

Herbert R. Harvey and Barbara J. Williams

INTRODUCTION

The Spanish conquest of Mexico in 1521 crushed the Aztec State and also signalled an end to four millennia of autochthonous development of Mesoamerican civilization. The century after the Conquest was demographically unsuccessful for native populations, so that knowledge of many aspects of Indian culture quickly disappeared. Much was already lost by the time the earliest chronicles describing native civilization were written. Piecing together the fragmentary record of Mesoamerican achievements has been a slow process. For example, it was not discovered until the late 19th century that the Maya had a mathematics based upon positional notation and the use of the zero.

Aztec civilization crowns a long line of cultural development in the Valley of Mexico. Large-scale engineering projects in and around their capital city of Tenochtitlan, in themselves, suggest that Aztec engineers utilized mathematics. The surviving written records of Aztec tribute reflect a complex system of counting expressed in numerous hieroglyphic symbols. Their State, built upon conquest and sustained by tribute payment from conquered provinces, extended to the southeast, well into the zone of Maya peoples and culture. Therefore, it is plausible that the Aztecs possessed a mathematical sophistication similar to that of the Maya, but the problem has been to uncover the direct evidence. The significance of our decipherment of two early post-Conquest census-cadastral documents from the Valley of Mexico is that

it provides, for the first time, direct empirical evidence
that Texcocan Aztecs used a positional notation system and
also developed a special symbol for zero.

The system of numerical notation most frequently associated with the Aztecs is described in the preceding paper by
Payne and Closs. However, there is an additional system,
employed by the Texcocan-Aztecs, which may be labeled as
'positional-line-and-dot'. This system was more efficient to
write, since it employed only four symbols which were easy to
draw: a vertical line, a bundle of five lines linked at the
top, a dot, a corn glyph (<u>cintli</u>), and position to indicate
the value of the symbols. At present we have noted its use
in land documents, but it would seem more widely applicable
and, moreover, well suited in general for arithmetical
calculations.

DOCUMENTARY EVIDENCE

Positional-line-and-dot notation is known from several
early post-conquest native Texcocan pictorial manuscripts.
The two most important are in library collections. One, the
Códice de Santa María Asunción, is in the Biblioteca Nacional
of Mexico. The other, the Codex Vergara, is in the Bibliothèque Nationale of Paris. We were especially interested in
these two native manuscripts because of their large corpus of
quantitative data and fine state of preservation. Also
intriguing were the large sections which had never been deciphered and the fact that the places mentioned in the codices
had never been specifically located. Thus, the challenge was
not only to decipher and analyze the contents of the codices,
but also to locate the place names. Beginning first with the
Códice de Santa María Asunción, we set out from Texcoco visi-
ting all nearby towns and barrios whose patron is Santa María

Asunción. Our search was finally rewarded in the village of
Tepetlaoztoc, which has a barrio called Asunción. The native
name of that barrio proved to be Cuauhtepoztla, the first and
most prominent town listed in the codex of Asunción. Moreover, when we explained the purpose of our questions, residents of that barrio led us to their church and showed us a
19th century document, which they now call their "títulos".
It was a copy of an early 17th century manuscript which
described the codex of Santa María Asunción. We knew then
that the origin of that codex was resolved. Subsequent field
work led to the conclusion that the communities in the Codex
Vergara have disappeared. However, our colleague Manuel
Arredondo found for us a late 16th century map in Mexico's
National Archive which aided us in locating several communities listed in the Codex Vergara. These were in another
barrio of Tepetlaoztoc. Thus, both codices pertain to this
village.

Each codex is bound as a book, the Asunción codex with 80
folios and the Vergara with 56. Each contains a census by
household and records household landholdings for a total of
16 named localities. The writing is in native style hieroglyphics characteristic of the Acolhua (Texcocan) kingdom of
the early 16th century. Both contain notations in script
that were added subsequent to the original drafting, including some late 16th century dates. The documents were signed
by Pedro Vásquez de Vergara, known to have been ordered by
colonial authorities to go to Tepetlaoztoc in 1543 to adjust
taxes. He appears again in litigation records of the 1550's
between the Indians and their encomendero, Gonzalo de Salazar
(Gibson 1964, pp.77-80). The fact that so many individuals
included on the census were deceased, as indicated by the
shading of their faces, together with several dated events on

the documents, suggests that they were prepared before the
onset of the great pestilence of 1545 to 1548, or circa 1545.
Their appeal, in other words, is that they were drawn only
one generation after the conquest.

Both codices are examples of cadastral records kept by
Indian communities in preconquest and early colonial times.
Alonso de Zorita, (1963, p.110) an astute mid-16th century
observer of native life, stated that:

"This principal is responsible for guarding and defending
the calpulli lands. He has pictures on which are shown all
the parcels, and the boundaries, and where and with whose
fields the lots meet, and who cultivates what field, and what
each one has... The Indians continually alter these pictures
according to the changes worked by time, and they understand
perfectly what these pictures show...".

We know that the Indians of Tepetlaoztoc specifically kept
such records because in 1551 the judge assigned to the litigation between the Indians and their encomendero ordered the
Indians to exhibit their paintings and explain their meaning
to him (AGI, Justicia, leg.151). Certainly their contents
were not any more self-evident then than they are today to
one unacquainted with the Indian conventions of writing and
notation. It is quite possible that both codices (or copies
thereof) were among the documents presented by the Indians,
since the documents had shortly before been officially certi-
fied by Pedro Vásquez de Vergara.

Each of the two codices is divided into three parts by
locality. The first section (tlacatlacuiloli, tlacanyotl)
contains a census by household, usually five households to a
page. The name for each head of household is written in
glyphic form beside the conventional symbol for household
head (fig. 9.1a). The second section (milcocoli) consists of

a record of land parcels associated with each household. The
term milcocoli may be a Náhuatl metaphor meaning field pattern. In modern Texcoco, certain small, rhombus-like cakes
are called cocoles, and if many of these were arranged on a
surface, they would create a lattice-like structure similar
to field patterns. In the milcocoli section, the scribe drew
the approximate shape of each field. The measurement of each
side was recorded using lines and dots, a line equal to one
linear unit, a dot equal to 20 (fig. 9.1b) (Cline 1966;
1968). The early 17th century native Texcocan historian,
Fernando de Alva Ixtlilxóchitl (1977, II, p.93), tells us
that in the Texcocan area the standard unit of linear measure, the quahuitl, was equal to 3 Spanish varas, or 2.5
meters. Units less than one quahuitl were indicated by
glyphic symbols such as a hand, an arrow, and a heart. The
modern equivalents of these glyphs can only be estimated at
present (Williams 1984, p.107; Castillo 1972, p.217). In
addition to recording linear measurements around the field
perimeters, each field also contains a glyph in the center
which indicates the type of soil (Williams 1980a; 1980b).
The significance of the milcocoli section for our purposes is
that it allows us to compute the area of each field and estimate the amount of land held by taxpaying commoner families
(macehualtin) in the early post-conquest period. Since the
two documents combined depict more than 1,000 land parcels,
they provide a record of landholdings unparalleled in the
area of Mesoamerica for such an early time period.

The third section of each locality contains another land
register labeled tlahuelmantli, which follows the same household sequence. There is little question that the tlahuelmantli section is a second register of the same lands. However,
the two land registers differ in several important aspects.

A

B

C

Fig. 9.1. Portions of the Códice de Santa María Asunción relating to the household of Pedro Tlacochquiauh. (a) The household census, tlacatlacuiloli, shows the head Pedro, his wife Juana, their two young daughters Ana and Martha, and the head's younger brother (teicauh), Juan Pantli, his wife María, and their son Balthasar (Parthasal). The shaded heads indicate that the individual depicted was deceased. (b) The milcocoli section records the approximate shape, perimeter measurements, and soil type of four fields belonging to the household, two to the household head and two to the head's brother. In the numerical notation employed in the milcocoli, lines equal 1 quahuitl and dots equal 20. The hand glyph shown in the first field indicates a fraction of a quahuitl. (c) The tlahuelmantli section shows the same fields as the milcocoli, but they are depicted as abstract rectangles, and a different numerical notation is used. In this positional-line-and-dot system, numbers occur in three registers. Lines in the upper right protuberance record units of 20. Lines on the bottom line of the rectangle (second register) or in the center (third register) are multiplied by 20. Dots, which occur only in the third register, equal 20^2 or 400. The corn glyph (cintli) at the top margin of the rectangle indicates zero in the third register. The numbers record the area of each field in square quahuitl. (Photographs from folios 2r, 10r, and 19v of the Códice de Santa María Asunción.)

The lands in the tlahuelmantli are shown in stylized form as rectangles of the same size. Perhaps not coincidently, the Nahuatl term **tlahuelmantli** literally means 'smoothed, leveled, or equalized', according to Siméon (1977, p.691). Also, the majority of these tlahuelmantli fields have a protuberance in the upper right hand corner (fig. 9.1c). In addition to the standardized field shapes, the placement of numbers in the tlahuelmantli is different than in the milcocoli. The numerical quantities using lines and dots are entered either in the center or on the bottom line of the rectangle and in the protuberance. When numbers (which never exceed 19) are entered on the bottom line, a cintli glyph occurs near the top border of the rectangle.[1] In addition, most fields contain a number ranging from 1 to 19 in the upper right hand corner protuberance. We have determined that these numbers report the area of the field in square quahuitl by use of positional notation. With this knowledge we may interpret the standard tlahuelmantli field form as a glyph meaning "area".

POSITIONAL NOTATION

Both the Códice de Santa María Asunción and the Codex Vergara have formed part of the known native pictorial record for over 150 years, and the latter was studied by such noted Mexicanist scholars as J.M.A. Aubin (1891) and Eduard Seler (1904). However, the codices have been little used because large sections remained undeciphered, notably the tlahuelmantli, which comprises one-third of each manuscript.

Our decipherment of the tlahuelmantli code came about by comparing areas calculated by us from the milcocoli dimensions with the Indian notations in the tlahuelmantli. We calculated milcocoli areas by two methods. Since fields are

not drawn to scale on the manuscripts, angles cannot be
directly ascertained. Thus, for quadrilaterals, given the
length of sides a, b, c, and d, we assumed a right angle
between sides a and b, which allowed calculation of the area
of the two resultant triangles. For non-quadrilaterals
(roughly 40 percent of the cases), field sides were drawn to
scale on graph paper and side angles were chosen so as to a)
produce field shapes consistent with shapes observable today,
b) maximize area, and c) preserve the relative shapes
recorded in the drawings. Area was then derived by counting
squares. Neither method accounted for side lengths in fractions of the standard quahuitl, which, where these occur,
would result in an understatement of the actual area.

Pursuing the hypothesis that the tlahuelmantli section
records a different set of data for the same fields illustrated in the milcocoli, we assumed that tlahuelmantli information might be related to field size, such as the amount of
seed sown or harvested, or the tax to be paid on each parcel.
We noted that in reading the tlahuelmantli numbers at face
value (i.e., dots for 20 and lines for 1), the numbers fell
consistently below those of the milcocoli areas in a ratio of
approximately 20 to 1. We found in hundreds of cases that by
multiplying the main tlahuelmantli number by 20, our area
figures derived from milcocoli calculations were closely
matched. In the process of testing this relationship further, it was discovered that in many cases our milcocoli area
figure would exactly match the tlahuelmantli number if the
lines in the upper right protuberance were treated as units
rather than as multiples of 20. Thus, it became apparent
that position of the numbers in the tlahuelmantli rectangles
had arithmetical significance.

The Texcocan positional-line-and-dot notation system

functions in the following manner. The tlahuelmantli rectangle records numbers in three positions, which we label
"registers". The first register, located in the upper right
protuberance, records the units of 20, indicated by 1 to 19
lines. Groups of 5 are bundled together by a connecting
line. The value of this register ranges from 0 to 19, and
when the number is zero, the protuberance is either not drawn
or is left blank. The bottom line of the rectangle constitutes the second register, and it expresses 1 to 19 units of
20 (i.e., vigesimal multiples from 20 to 380). To derive the
total value, the number in the second register is multiplied
by 20 and then added to the number in the first register.
The sum of the two registers never exceeds 399. The central
portion of the rectangle constitutes the third register, and
expresses quantities of 400 or greater in multiples of 20.
Again, to derive the total, the number in the third register
is multiplied by 20 and added to the quantity entered into
the first register. The Texcocan symbol for 20 (the dot)
only occurs in the third register, and according to its positional value means not 20 but 20^2 or 400. The second and
third registers are never used concurrently. When there is
no entry in the third register, the cintli glyph is drawn
toward the top of the rectangle and signifies zero in the
third register. Thus, for the first field illustrated in
figure 9.1c the entries are one dot and 11 lines in the third
register $[20^2 + (11 \times 20)] = 620$ plus 4 lines in the first
register, which totals 624 square quahuitl. The third field
has an area of 333 square quahuitl as recorded by 16 lines in
the second register (16 × 20 = 320) plus 13 lines in the
first register, and a cintli glyph at the top of the rectangle indicating zero in the third register.

AREA COMPUTATION

The milcocoli register depicting linear dimensions of
individual field boundaries confirms evidence from other
native documents and early Spanish descriptions that native
Mexicans maintained detailed cadastral records. However,
decipherment of the tlahuelmantli adds to this knowledge the
hitherto unknown practice of expressing landholdings in terms
of area. Thus, it also provides an insight into the practical application of native arithmetic apart from its esoteric
use in calendrics and astronomy.

Just how area was determined remains to be resolved. The
milcocoli register does not provide enough information to
calculate exact area of fields, particularly the highly complicated forms, since it does not record the true shape by
means of angles or auxiliary measures. In other words, the
milcocoli could not have been used as a "worksheet" for the
tlahuelmantli section. Therefore, there must have been
intermediate steps in area calculation between the milcocoli
and tlahuelmantli records. One method could have employed a
grid system in the field at the time of survey; another could
have combined a grid system with computation; a third could
have recorded in some fashion the required information for
later computation of area. The key to discovering the most
likely method rests potentially in the documents themselves.

To explore how area was determined and the degree of accu-
racy achieved, we examined quadrilaterals from the locality
called Cuauhtepoztla. A preliminary analysis indicates that
55 percent of the tlahuelmantli areas fall within 5 percent
of our areas calculated from milcocoli data; and 71 percent
of the tlahuelmantli areas fall within 10 percent of our figures. Therefore, there is considerable concordance between
our figures and theirs. More specifically, 8 percent of our

area calculations based on milcocoli data and Texcocan calculations as expressed in the corresponding tlahuelmantli
statements correspond exactly. Most of these cases involve
quadrilaterals whose opposite sides are equal. We assumed
these to be squares and rectangles for which the length
times width algorithm applies. Since the Texcocan figures
correspond to ours, this suggests that they also used the
same algorithm. In 26 percent of the cases, the tlahuelmantli areas fall below our estimates, but the Texcocan figures
may be more accurate than ours because, if our right angle
assumption did not approximate the actual field form, then
our method over-estimated the area. On the other hand, for
66 percent of the cases the tlahuelmantli areas are greater
than our figures. This is puzzling because our right angle
assumption approximates the maximum area possible given the
side lengths reported in the milcocoli (Harvey and Williams
1980, pp.500-501).

Some of the discrepancies are explained by two additional
algorithms. The tlahuelmantli area of many quadrilaterals
which are nearly parallelograms can be obtained exactly by
taking the area to be the sum of two right triangles whose
legs are the sides of the fields. For more irregularly
shaped quadrilaterals, we can frequently derive the tlahuelmantli estimate by summing the area of the largest possible
rectangle within the field and the area of the remaining
triangles. Thus, the data strongly suggest that Aztec surveyors used some sort of arithmetic calculation to derive
area of fields surveyed.

Not all of the differences between our area calculations
and the tlahuelmantli figures can be accounted for by different methods of computation. Internal evidence in the two
codices suggests that the milcocoli and the tlahuelmantli

registers were recorded at different times and perhaps by different surveyors. For example, the sequence of fields possessed by each household is not invariably the same in the two registers. Also, some individual households have added or lost parcels between the two registers. Therefore, it appears that the two registers were either based on separate surveys or one is an updated version of the other. Insofar as they may represent partially or entirely separate surveys, some discrepancies may be expected in the linear dimensions of the fields due to survey inconsistency. A time lapse between surveys might also have resulted in a real change in field size, such as the result of subdivisions or consolidation due to inheritance or property exchange. The larger discrepancies probably relate to such realignments.

It is noteworthy that when all (approximately 1100) individual field areas are aggregated at the household and locality level, the discrepancies between our estimates and the tlahuelmantli figures tend to even out. At the household level, in some cases large milcocoli fields are replaced by small fields in the tlahuelmantli and vice versa with the result that household aggregate landholdings remain fairly constant. At the locality level, the loss by one household is another's gain, so that aggregate locality totals show slight to negligible variation (Harvey and Williams 1980, p.503). Overall this implies not only reasonably accurate measuring and careful record keeping, but also methods to derive area that yielded reasonably accurate results.

APPLICATION

The functional significance of calculating land area is most apparent in relation to the system of taxation in ancient Mexico. Landholding commoners paid property taxes

(tribute) based on the size and quality of their lands.
Hernán Cortés (1538, p.542) reported that "...he who has them
can pay tribute because for each measure [our emphasis] so
much tribute is charged them according to where the lands are
located." His son, Martín Cortés (1563, p.443) noted that
"...he who has a piece of land paid a tribute; and the one
with two, two; and the one with three, three; and he who had
a piece of irrigated land, paid double that of one who had
dry land." Central Mexico's topography is such that much of
the taxed land had to have been located in hilly or mountainous terrain, and square or rectangular fields the exception
rather than the rule. A tax system based on standard "measures" would have required a mechanism for equating oddly
shaped fields with regular ones. This was the task of the
surveyor, and as can be seen from the two Tepetlaoztoc codices, even very eccentrically shaped land parcels could be
expressed accurately in terms of their square unit content.
The importance of survey in Aztec society is reflected in
16th century Nahua vocabulary, which included numerous terms
related to surveying, and interestingly, several referring to
incompetent surveyors (Lameiras 1974).

There is increasing evidence that the basic areal
"measure" of land (analogous to hectare) was 400 square units
(20 × 20 linear units). In the case of Tepetlaoztoc, the 400
square measure was equivalent to .25 hectares
    $[(20 \text{ quahuitl} \times 2.5 \text{ meters})^2]$ = 2500 square meters.
Air photos of Tepetlaoztoc show small relic fields identifi-
able by soil discoloration in an area where fields recently
have been enlarged for mechanized agriculture. These were
square fields that measure 400 square quahuitl. In other
areas of the Valley of Mexico the length of the linear unit
varied, but the concept of a standard measure 20 by 20 square

appears to have prevailed. Many extant pictorial documents from Central Mexico depict land parcels as rectangles and with dimensions or quantities expressed as multiples of 20. The Tepetlaoztoc codices suggest that some of the fields depicted in other documents might not be actual parcels, but tlahuelmantli-type abstractions. Precise shape or form of a field was less relevant for tax purposes than the size.

The Codex Mariano Jimenez (1967; also Leander 1967) offers a case in point. Folios 6 and 7 of that document depict a list of 11 rectangular fields, each 20 brazas in width, but increasing in length in regular increments from 100 to 800 brazas (fig. 9.2). They are ordered on the pages according to length, and tribute to be paid is noted beside each field. Of the three types of tribute assessed, two (firewood and turkeys) are fixed quantities regardless of field size; the third, cacao beans or coins, varies in direct proportion to the size of the field. For example, the tribute for a 400 square braza field is 20 cacao beans, indicating a tax rate of one cacao bean per 20 square brazas. These fields have been interpreted in the past as actual field dimensions of commoner landholdings in Otlazpan. An alternative interpretation is warranted because perfectly rectangular fields seem hardly possible, given the terrain of the Otlazpan area (modern Tepeji del Río), and especially considering that roads, ditches, trees, fences, and constructions make regular shaped fields unlikely even on lacustrine plains. Furthermore, the largest area shown on the Otlazpan document is 16,000 square brazas, which would have been extraordinarily large for a 16th century field possessed by a commoner. In Tepetlaoztoc the largest fields rarely exceeded 1000 square units, although in the aggregate many families possessed over 5000 square units. In our view, the Otlazpan field list appears to represent standardized abstractions of aggregate

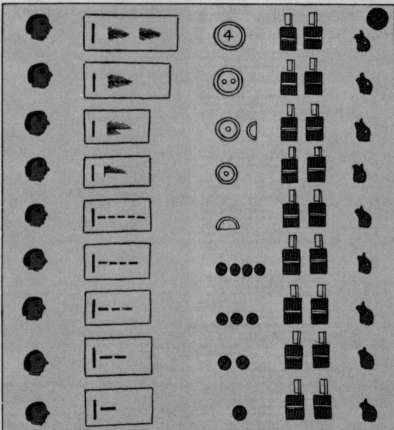

lo ga se tributan cada macegual (IИ) pueblo conforme ala canti das de
tierra q̃ tiene y posee el q̃ tuviere ochocientas bracas desementera delargo y veyn
te en anc̃ a se tributan cada ochenta dias quatro Reales deplata y quarenta
ffaxas deleña y cada año una galllina dela tra / y el q̃ tuviere quattocientas
bracas desementera delargo y veynte en anc̃. a de tributan cada ochenta dias
dos tomines deoro comun y quarenta ffaxas deleña y cada año una galllina
y el q̃ tuviere trezis bracas desementera delargo y veynte en anc̃. a de tribu
tar cada ochenta dias un tomin um y quarenta ffaxas deleña y cada año
una galllina dela tra. y el q̃ tuviere dozis bracas delargo y veynte en anc̃.

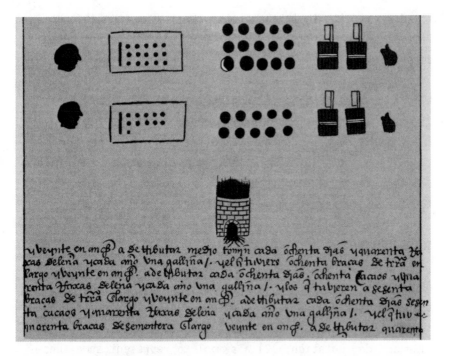

Fig. 9.2. The Codex Mariano Jimenez (1967) shows eleven rectangular fields, each 20 brazas in width, but with lengths varying in regular increments from 100 to 800 brazas. In our view, the drawings are not of actual fields with these dimensions. Rather, the numbers indicate area in square brazas. Instead of indicating area measures directly, as in the Vergara and Asunción codices, the Otlazpan scribe assumed a constant multiplier (20 brazas) and a variable multiplicand, whose product would equal the area measure. For example, landholdings equal to 8000 square brazas were indicated by depicting a rectangle 20 brazas wide by 400 brazas long. To the right of the list of land areas is the tribute levy, so that a landholder could determine the tribute on his holdings.

field areas with their associated tax levy. As such, the
list functioned as a "tax table" from which a person could
determine the tax on his holdings. If the Otlazpan tax rate
of one cacao bean per 20 square units were applied to Tepetlaoztoc, then taxes on each parcel could also be read
directly from the tlahuelmantli second and third registers by
reading the numbers at face value rather than as multiples of
20. Thus, a tlahuelmantli field with two dots in the third
register would be read as 800 square quahuitl in area, and
the tax would have been 40 cacao beans.

CONCLUSIONS

As a result of the decipherment and preliminary analysis
of two early post-conquest pictorial manuscripts from Tepetlaoztoc, three new features emerge concerning the native
arithmetical system of the Texcocan Aztecs: 1) positional-
line-and-dot notation; 2) a symbol to represent some functions of zero; 3) a probable set of algorithms to compute
land area. However, since both documents were drafted in the
1540's we must ask whether the system was prehispanic in use
and development or whether it might have been introduced by
the Spaniards.

There are a number of features associated with Spanish
mathematics which strongly contrast with the Texcocan-Aztec
system. Spanish mathematics of the early 16th century used
base 10, although some weights and measures were expressed in
other bases such as 8 and 12, which derived from earlier conventions in the Mediterranean area. The Arabic system of
notation had been adopted in Spain prior to the 16th century,
but earlier Roman notation continued to be used extensively
until the end of the 17th century. In both Arabic and Roman
notation, horizontal but not vertical place value is of

fundamental significance. In contrast, in Texcocan notation, horizontal position could be ignored so that, for example, the number 23 could be written either as ///• or •///, but vertical position was fundamental to reading the proper value of the number. The zero was highly developed in Arabic arithmetic but only weakly in Texcocan-Aztec. Finally, the Texcocan system was internally consistent throughout and reflected the idiosyncracies of the spoken language. The weight of evidence strongly supports an indigenous development of Texcocan-Aztec arithmetic rather than a post-conquest adaptation to new ideas. The absence of description of the native arithmetical system in the early chronicles suggests that Spanish observers of native culture, so astute in their description of some things, were either ignorant of native arithmetical practice or unimpressed (Morley 1915).

It is, perhaps, in the practical application of an arithmetical system to derive area that Spanish and native practices most sharply contrast. The Spanish could but usually did not express land units in square measures. The size of agricultural plots was frequently designated in terms of yield, such as so many fanegas of wheat. Also, Spanish land units such as the caballería were notoriously variable. Traditionally, the Spanish had been content with imprecise descriptions of land parcels, whereas the native Mexicans had not. This contrast suggests the continuation of prehispanic practices, rather than an adaptation to a Spanish system. As Anderson, Berdan and Lockhart (1976, p.5) observed "...from the mid-sixteenth century, as far back as our selections go, central Mexican Indians were measuring their lands very exactly, down to the yard in both dimensions, using quite sophisticated and individual terminology. At that time Spaniards in Mexico were still transferring land by the league, with no other description than the names of nearby owners or outstanding geographical features."

Three basic concepts of the Texcocan positional-line-and-dot system are shared with the Maya: base 20, vertical position, and the use of zero. As in Maya arithmetic (Lizardi Ramos 1962; Fulton 1948), the zero concept in Texcoco appears to have undergone only a limited development compared to functions and significance of the zero in Arabic mathematics. Since Aztec peoples were relative late-comers to the Valley of Mexico and the zone of high civilization, and since their area of political and economic control eventually extended to the borders of Maya-speaking peoples, the arithmetical system used in the Texcocan area may reflect a direct borrowing from the Maya. However, Texcocan-Aztec arithmetic was more likely a regional expression of a basic set of conventions and principles known throughout the Mesoamerican area for two millennia or more.

Ciphers expressed by picture symbols were geographically widespread in Mesoamerica, but 16th century pictorial documents from Central Mexico show that a number of other more abstract symbols were also used. In the Otlazpan manuscript, both a flag and a bar were employed as alternative symbols for 20. In Tenochtitlan, bars and dots, reminiscent of the Maya, were symbols used in calendrics. The Codex Kingsborough, also from Tepetlaoztoc, extensively employed picture symbols, in many cases combined with lines and dots (Paso y Troncoso 1912). While the choice of symbol to record numbers may have been the scribe's, it seems more likely that the determinant was the object to be counted and that the numerical classifier system of the spoken language was reflected in writing. This could explain why picture symbols appear in one context and abstract symbols in another.

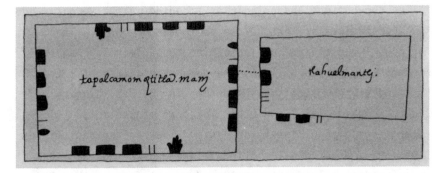

Fig. 9.3. The "tlahuelmantli convention", which indicates areal extent of a field by positional-line-and-dot notation, was utilized in Texcocan land documents other than the Vergara and Asunción codices. Shown above are drawings of a field called Tapalcamomoztitla from a document in the Papeles de la Embajada Americana Collection of the Archivo Histórico of the Museo Nacional de Antropología in Mexico City. The field on the left records data drawn in the milcocoli convention and the other linked to it by a dotted line is labeled tlahuelmantli. The area of the field shown in the milcocoli convention may be roughly calculated (omitting fractions of a quahuitl) by multiplying length (17 quahuitl) by width (16 quahuitl). The area is thus 272 square quahuitl. The number depicted in the tlahuelmantli convention is also 272. Thirteen lines on the inside left margin are multiplied by 20 (=260) and the 12 lines on the outside bottom line are multiplied by 1 (=12) totaling 272. (Archivo Histórico del INAH, leg.7, exp.2-24.)

The conventions and principles embodied in the Vergara and Asunción codices are not unique to those documents. The "milcocoli convention" of indicating measurements around the perimeter of fields by lines and dots with values of 1 and 20 is found on the Oztoticpac Lands Map in the U.S. Library of Congress (Cline 1966) and other Texcocan manuscripts. Similarly, the "tlahuelmantli convention" showing abstract fields with area indicated by positional-line-and-dot notation has been found on a late 16th century document from another Texcocan locality (fig. 9.3) (Papeles de la Embajada Americana n.d.). Further, the cintli glyph appears in a context other than tlahuelmantli to clarify place value. Since the direct evidence we have for the positional-line-and-dot notational system is thus far restricted to the Texcocan province of Acolhuacan, and largely to land documents within this area, it may well have been a special system devised for the purpose of land description. On the other hand, it is a system in which arithmetical computations could be accomplished with clarity and facility in contrast to the more elaborate picture symbols commonly used in tribute records. Until the Tepetlaoztoc documents were deciphered, use of positional notation, the zero symbol, and derivation of land area by Aztec peoples were unsuspected. The significance of the Tepetlaoztoc decipherment is that many documents with hitherto undeciphered numerical notations may now be re-examined and our understanding of prehispanic arithmetical practices possibly greatly expanded.

NOTE

1. The glyph depicts a shelled ear of corn, which usually is read olotl. However, the same glyph appears in name glyphs in these codices with the phonetic value of çi or çe.

Thus, we read <u>cintli</u>, ear of corn.

ACKNOWLEDGEMENTS

This research was supported by NSF grant BNS 7725659 (to B.J.W.) and NSF grant BNS 8111804 (to H.R.H.). We thank the Centro Internacional de Mejoramiento de Maíz y Trigo for providing facilities in the field zone and <u>La Recherche</u> for permission to publish this modified version, in English, of the article "L'arithmétique azteque" (Harvey and Williams 1981).

10. Mathematical Ideas of the Incas

Marcia Ascher

INTRODUCTION

In the first third of the sixteenth century, Europeans
reached the Andes and "discovered" a culture known as the
Incas. Within about thirty years, the culture was destroyed.
When we talk about the Incas, we are discussing a group that
inhabited a particular geographic area; shared a common language; had a particular social, political, and economic
organization; and existed during a particular time period.
The region they inhabited is described today as all of the
country of Peru and portions of Ecuador, Bolivia, Chile, and
Argentina. It is a region that has coastal deserts as well
as mountainous highlands. There were many different groups
in this region. But, starting about 1400, one of the groups,
the Incas, moved slowly and steadily upon the others. The
groups retained many of their individual traditions but were
forcibly associated into a bureaucratic entity. Using and
extending different aspects of the cultures of the different
groups, the association was accomplished by the spread of one
common language, Quechua; the building of an extensive road
system; the overlay of a common state religion; the imposi-
tion of a careful system of taxation; the broad extension of
irrigation systems; and the building and utilization of
storehouses for the methodical storage and redistribution of
agricultural products. The Inca, as we use the term, refers
to this extensive organization of three to five million people which existed from about 1400 to 1560. The Inca were
what is usually called a civilization except that they did

not have writing in the sense we know it. Thus, there are
not, and cannot be, any discursive records in their own
words. The Incas did, however, have an unusual logical-
numerical recording system of quipus which will be elaborated
later.

The Incas no longer exist and information about them is
fragmentary. There are basically three sources that are
used: their artifacts that still remain; the writings in
Spanish by contemporaries of the Incas; and studies of modern
Quechua speakers who have continued to inhabit the same
region. To find out, in particular, about the mathematical
ideas of the Incas, each of these sources presents problems.
Artifacts are concrete objects. They result from ideas and,
at best, with caution, some of the ideas themselves can be
inferred. The chronicles, written by the Spaniards who
entered the Andes, contribute to scholarly study and knowledge about the Incas but cannot be taken simply and directly
as stated. They are writings by people who lived in the sixteenth century, historically related to us but from a culture
different than our own, about people who were very different
from themselves whose culture and ideas they did not necessarily understand and certainly did not appreciate. Ethnographic studies of modern Quechua speakers also have severe
limitations. Any culture, in all its ramifications, comprises a whole. Changes in concepts accompany changes in
political, social, and economic systems; changes in skills;
and changes in artifacts. Even when some similar artifacts
are used, they are not necessarily embedded in the same ide-
ational context. However, relying primarily on archeological
artifacts, since they are the only source that is actually
Inca, something can be known of the spatial, numerical, and
logical ideas of the Incas. And, the ethnohistorical studies

can provide more general background information.

NUMBER WORDS

The sixteenth-century chroniclers and the contemporary Quechua language indicate a base ten number system. The language has distinct words for 1 through 10, 100, 1000, and 1,000,000. A number word placed before the word for one of the powers of ten and smaller in value than it is a multiplier; placed after one of them, it is additive. (The additive number word takes a suffix of yoq which, when it follows a consonant ending, takes a connective of ni.) The words for 10, 100, and 1000 are chunka, pachaj, and waranqa; and for 3 and 8, they are kinsa and pusaq. Then, for example, 38 is kinsa chunka pusaqniyoq; 83 is pusaq chunka kinsayoq; 838 is pusaq pachaj kinsa chunka pusaqniyoq; and 3800 is kinsa waranqa pusaq pachajniyoq (Manya A. 1972, p.13; Bills, Vallejo and Troike 1969, pp.67-68).

The construction of number words and the symbolic representation of numbers are distinctly different concepts. We, for example, say "seventy eight" whether we read the symbols LXXVIII or 78 and the French write 78 even when calling it "soixante dix-huit." From the quipus of the Incas, however, it can be seen that their symbolic representation of numbers also used a base ten positional system. (The specifics of the symbolic representation will be elaborated in the discussion of the quipu.)

CALENDRICS

The calendric ideas of a culture are a part of, or related to, their mathematical ideas. The development of a calendar, on the one hand, depends on scientific observation of the physical world and, on the other hand, imposes a structure on

the passage of time. Also, through a calendric system,
social, political, or religious aspects of a culture are correlated with periodic natural phenomena.

A great deal of work is being done on the Inca calendar by
R.T. Zuidema (1964; 1977; 1982) and his associates. His primary focus is the ceque system of Cuzco. The ceque system
refers to 328 huacas, "sacred shrines", arranged in groups of
3 to 13 along 41 lines (the ceques) radiating from Cuzco. He
believes that the ceque system was partially used to observe
astronomical events and, also, that within its arrangement is
a record of the arbitrary time units of the Incas as well as
the astronimical time units that were of concern to them
(Zuidema 1977). From his analysis of the shrine and ceque
arrangements and related historical material, he concludes
that the Inca had a complex system involving several cycles.
They probably recognized 365 days as the solar year, 584 days
as the Venus year, and 657 days as a double sidereal lunar
year. They, perhaps, simultaneously used synodical months of
29 or 30 days, a solar year of 12 months of 30 days or
longer, a lunar year of 12 sidereal lunar months of 27 and
1/3 days each, and months of 23, 24, or 26 days, as well as
weeks of 8 days and weeks of 10 days.

While, at the present moment, there is no conclusive specific description of the Inca calendar, study continues on
the ceque system and the associated astronomical ideas. Calendric information would, no doubt, have been among the
things recorded on quipus.

A COUNTING BOARD (?)

A 1179 page letter about the Incas, written to the Spanish
king in about 1600, contains 397 drawings. In the corner of
one drawing, there is a 5 row by 4 column grid. The squares

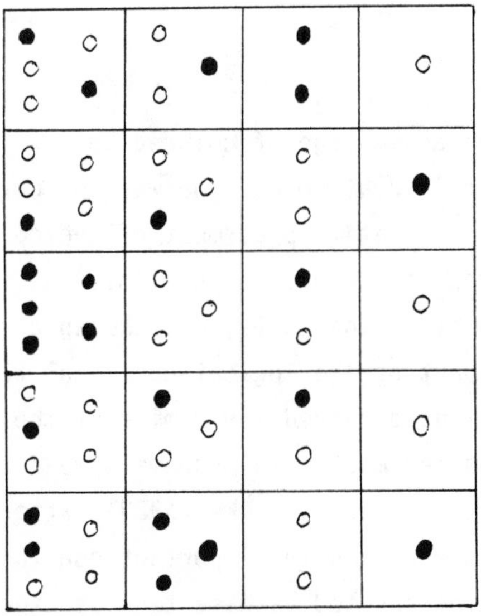

Fig. 10.1. An Incan counting board (?).

of the grid contain 5, 3, 2, and 1 small circles arranged
uniformly in each of the first through fourth columns respectively. Some of the circles are filled in and some are empty
(fig. 10.1). Because it is associated with the figure of a
man holding a quipu, it has been argued that the object is a
counting board. There is, however, no other evidence for its
use. There are various interpretations of how a base ten
number could be read from the configuration (Ascher 1972; Day
1967; Locke 1932; Wassén 1931). Alternatively, Zuidema
(1977) suggests that since there are fifty-five circles on
the grid, each is associated with a day in a double sidereal
month or with a particular huaca or ceque. In Ascher and
Ascher (1980), the letter and its various drawings are discussed at greater length and it is suggested that the picture
not necessarily be taken so specifically. It might not be
realistic in detail and it might not even depict an object

used by the Incas.

QUIPUS

A quipu is an assemblage of colored knotted cords. The colors of the individual cords, the way the cords are connected together, the relative placement of the cords, the spaces between the cords, the knots on the individual cords, the relative placement of the knots, and the spaces between the knots, are all part of the logical-numerical recording.

About 550 quipus survived the demise of the Inca state and these are now housed mostly in museums spread throughout the world. In the 1920's, L.L. Locke (1923), after examining about 50 of them, reached the important conclusion that the knots could be interpreted as base ten positional numbers. While this is true of many of them, there is more to quipus than that. About a half dozen people studied about forty more quipus in the next forty years. More recently, an exhaustive study of quipus was undertaken. About 450 quipus were examined and uniform detailed physical descriptions of 191 quipus are now available to enable others to further study them (Ascher and Ascher 1978).[1] (With these descriptions is a bibliography of previous writings about quipus and also a list of the current whereabouts of any known quipus.) What follows is based on our analysis of about 250 quipus. A fuller discussion of the cultural context and logical-numerical system of the quipus can be found in Ascher and Ascher (1980).

In general, a quipu has one cord, called the main cord, which is thicker than the rest and from which other cords are suspended. When the main cord is laid horizontally on a flat table, most of the suspended cords fall in one direction ("downward"). These are called pendant cords. Sometimes,

some of the suspended cords fall in the other direction
("upward") and so are called top cords. Suspended from some
or all of the pendant or top cords are other cords called
subsidiary cords. These can have cords suspended from them
so that there can be subsidiaries of subsidiaries and subsidiaries of them and so on. Sometimes there is a single cord
attached to the end of the main cord. Since the way it is
attached is different than the attachment of a pendant or top
cord, it is referred to as a dangle end cord. All attachments are tight so that cord positions relative to each other
are fixed. Larger spaces between some adjacent cords sometimes set off groups of cords from each other. Pendant
cords, top cords, and subsidiary cords are about 20 to 50 cm
long. A quipu can be made up of as few as three cords or as
many as two thousand cords and can have some or all of the
types of cords described. A schematic of a quipu is shown in
figure 10.2.

COLOR AND ARRAYS ON QUIPUS
     Cord color is an important part of the logic of the cords
formed into a quipu. Colors primarily indicate cord associations or differences within the context of a single quipu.
The cord color order can, for example, be patterned so that a
sequence of, say, five colors is repeated four times. Each
cord can then be identified as the ith element in the jth
group where $i = 1, ..., 5$ and $j = 1, ..., 4$. When cord
placement and spacing are combined with color coding, a flexible but clear scheme emerges for identifying categorization
or cross-categorization. Consider, for example, a quipu with
sixty pendant cords spaced along the main cord such that they
are grouped into three sets of twenty each. If, in addition,
each set of twenty is a sequence of five colors repeated four

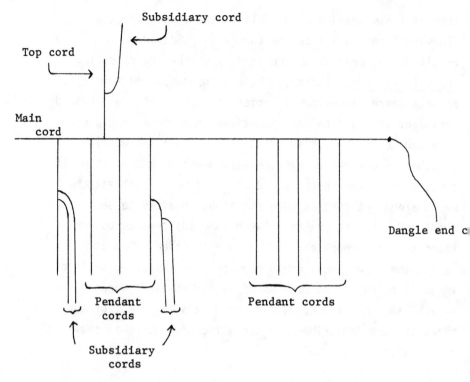

Fig. 10.2. A schematic of a quipu.

times, the sixty pendants form an array that can be described as p_{ijk} where i = 1, ..., 5, j = 1, ..., 4, and k = 1, 2, 3. This could be further complicated by having two subsidiaries, say red and blue, on each pendant cord. Distinguishing each subsidiary by the pendant to which it is attached and by its color, the one hundred and eighty subsidiaries then can be described as s_{ijkm} where i = 1, ..., 5, j = 1, ..., 4, k = 1, 2, 3, and m = 1, 2.

Just as pendants can be spaced along a main cord, subsidiaries can be spaced along a pendant cord. What is more, because they are subsidiaries of a particular cord, and because they too can have subsidiaries, another classificatory relationship is formed, namely, a hierarchical relationship. To visualize this relationship, a pendant and its associated subsidiaries can be thought of as a tree diagram. A set of pendants which all have the same subsidiary arrangement would then be analogous to a set of tree diagrams, all with the same number of levels and the same branches on each level. Thus, through subsidiaries, the recording of hierarchical relationships can be combined with the recording of categories and cross-categories. The resultant configuration increases in complexity as the number of subsidiary levels or the number of branches on the levels increase. For example, one quipu has pendants forming an array that can be described as a_{ij} where i = 1, ..., 8 and j = 1, ..., 10. From some of the pendants, however, there are as many as five levels of subsidiaries (i.e., subsidiaries, subsidiaries of subsidiaries, subsidiaries of subsidiaries of subsidiaries, etc.) and as many as ten subsidiaries on some levels. An example of a more modest configuration is a quipu with six pendants each of which has six first-level subsidiaries.

Different quipus can be seen to be in different stages of

completion. From them, it can be concluded that before any knots were placed on the individual cords, a blank quipu was completely formed. In the case of the tree diagram analogy, it would be as if one were to draw all the branches of the tree before entering any data. Or, as with a matrix, one would first make explicit that it is, say, an 8 × 7 matrix before entering any of the elements. It is, always, the structure which is provided for data, rather than the data itself, that carries the logic of the relationships between the data items. And, of course, data is only meaningful when viewed within the context of its logical structure.

On many quipus, color coding and cord placement are used to reinforce each other. For example, if twenty pendants were spaced along a main cord so that they were grouped into five sets of four pendants each and, if each of the five sets contained the same four color sequence, the description of the array would be a_{ij} where $i = 1, \ldots, 5$ and $j = 1, \ldots, 4$. However, while i corresponds to the set, j corresponds to either position in set or color. This apparent redundancy has some very important ramifications. It enables the exclusion of an unapplicable category with no ambiguity. If, in our 5 × 4 example, one of the sets contained only three pendants, their color would make clear which one had been omitted. Color coding, in addition, is used to reinforce subsidiary placement. This is analogous to drawing a set of tree diagrams so that the different branches on any given level are drawn in a particular order and particular color sequence. When a branch is unnecessary on one of the trees, it can be omitted without ambiguity making the tree more compact. When color coding is used for positional reinforcement, the color sequence of each cord grouping becomes a subsequence of the whole.

DATA REPRESENTATION ON QUIPUS

On the individual cords, whether they be pendants, top cords, subsidiaries, or dangle end cords, there appear only three types of knots: single knots (simple overhand knots), long knots (made up of two or more turns), and figure-eight knots. Knots are clustered together and separated by space from adjacent clusters. Basically, the knot clusters represent digits and each cord can be interpreted as one item of data formed from digits. The data item can be a number, multiple numbers, or a number label.

Where the knots on a cord represent a number, it is a base ten positional integer. Each consecutive cluster position, moving from the free end of a cord to where it is attached to another cord, is one higher power of ten. The units position has long knots and the other positions have clusters of single knots. Because of the way long knots are constructed, there cannot be a long knot of one turn and so a figure-eight knot is used for a one in the units position. The largest number seen on a quipu was a five-digit number, 97,357.

The concept of zero is of particular importance in a positional system. However, in its entirety, the concept has three aspects:

a) a position containing "nothing" contributes to the overall value of a number;

b) "nothing", in and of itself, is a number; and

c) there is a special symbol for "nothing".

Although on quipus there is no special symbol for zero, the other two aspects of the concept are present. Even without a special symbol, there is little ambiguity. One of the main reasons is that knot type distinguishes the units position from other positions. Also, the knot clusters are carefully aligned from cord to cord. Thus, a cluster position with no

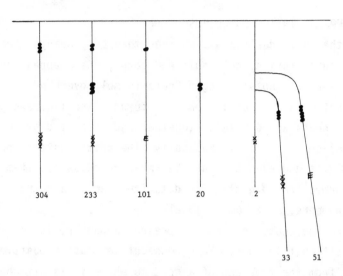

Fig. 10.3. Base ten positional numbers represented by knots (• = single knot; x = one turn of a long knot; E = figure-eight knot).

knots is apparent when related to other cords. Figure 10.3 is a schematic with some examples of numbers represented by knots. Ambiguity could arise in the case where all the numbers on a set of cords lack knot clusters in the same position as there is then no indication that such a position exists. (In figure 10.3, for example, this would be the case if the values were 3004, 2033, 1001, 20, 2. However, one would then expect an observably larger space between the top and middle clusters as compared to the middle and last clusters.) Where clusters are not carefully aligned, it is a problem of "handwriting" -- a problem that is encountered in any representation. There is also the question of a cord with no knots at all. On a quipu where color coding reinforces cord position, as discussed before, there is no ambiguity between a zero-valued cord and a blank cord because inapplicable categories and, hence, items left blank, do not appear. We cannot, however, distinguish zero from a cord not

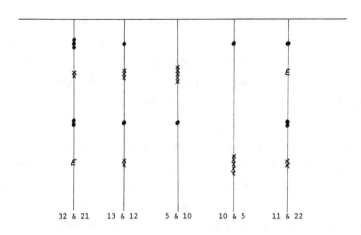

32 & 21   13 & 12   5 & 10   10 & 5   11 & 22

Fig. 10.4. Multiple numbers represented by knots (• = single knot; x = one turn of a long knot; E = figure-eight knot).

yet completed. This is a problem introduced by the fact that we are viewing an archeological artifact rather than a document in its proper situational context.

Where the knots on a cord represent multiple numbers, they are still base ten positional integers. The difference is, reading from the free end of the cord, each long knot (or figure-eight knot) is the units position of a new number. Figure 10.4 is a schematic with some examples of multiple numbers represented by knots. A maximum of three numbers has been seen on individual quipu cords.

The third type of data, while still numerical, are number labels rather than magnitudes. The use of number labels is quite common in our own experience. We write, for example, 1-607-273-5241 to represent the digits to be dialed for a long distance station-to-station call with 607 as the area code, 273 as the local exchange, and 5241 as the local phone identifier. We identify individuals by social security

numbers or employee numbers, products by product codes, and
automobiles by license plate numbers and engine serial numbers. It is of importance to recognize that quipus contain
numbers used in this way because, as a result, the quipu is a
far more general recording device than was previously
believed.

Numbers and number labels combined with logical structures
make the quipu considerably more complex and sophisticated
than the 1-1 knot records or mnemonic devices with which they
are so often misassociated in mathematical literature. They
differ from these also in that they constitute the sole
recording system used by a state which carefully planned and
executed large projects involving both people and natural
resources. While quipu construction and interpretation was
limited to a special but important class of people in the
Inca Empire, quipus were a universal system rather than a
personal ad hoc device.

How can one be sure of the numerical interpretation of
knots and how can the different data types be told apart? As
was noted before, data has to be viewed in the context of its
logical structure. It is found that when the knots are
numerically interpreted, there are corroborating relationships that are consistent within the logical structures of
the quipus. For a particular quipu, where arithmetic relationships are found, it can be inferred that the data are
numbers and not number labels. From other relationships,
number labels can be inferred. Unfortunately, in many cases,
no inference about data type can be made.

Cord arrangements, differential coloring, and knots are
visible in figures 10.5 through 10.7. Figure 10.5 shows an
extended quipu; figure 10.6 contains seven small quipus found
tied together; and figure 10.7 is a closeup of a quipu.[2]

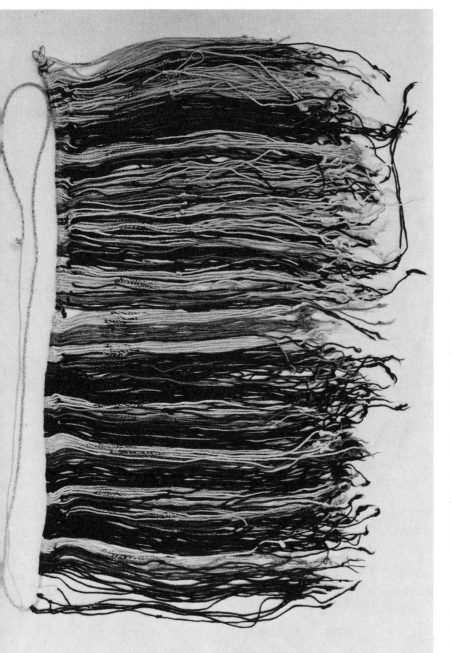

Fig. 10.5. An extended quipu from a private collection in Chile.

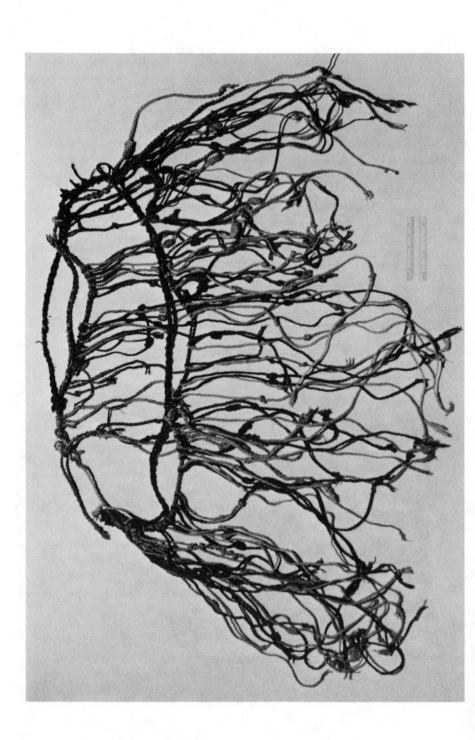

Fig. 10.7. A quipu from the Museo Nacional de Antropología y Arqueología in Lima, Peru.

ARITHMETIC AND STATISTICAL QUIPUS

On many quipus, numbers are found that are the sums of other numbers on the quipu, and this arithmetic relationship is present systematically. Where, for example, an array can be described as a_{ij} where $i = 1, \ldots, N$ (associated with spatial groups) and $j = 1, \ldots, M$ (associated with position in group or color), the sum of the numbers in the ith spatial group, namely, $s_i = a_{i1} + a_{i2} + \cdots + a_{iM} = \sum_{j=1}^{M} a_{ij}$ is found associated with the corresponding group on a pendant just before or after the group, or on a top cord in the middle of it. Top cords almost exclusively carry the sums of the values of the groups with which they are associated. On the other hand, if the sum of the numbers of the jth position in each group (or the jth color in each group), that is

$s_j = a_{1j} + a_{2j} + \cdots + a_{Nj} = \sum_{i=1}^{N} a_{ij}$ is recorded, it is found forming an $(N+1)^{th}$ group. This is carried even further so that some groups carry the sums of groups which in turn carry the sums of other groups. Three actual quipu examples of this summation over category are included to convey the idea more concretely.

Example 1:   AS199 has twelve cord groups, each with six pendants and a top cord, thus forming an array a_{ij} where $i = 1, \ldots, 12$ and $j = 1, \ldots, 6, T.^3$ On it, for $i = 1, \ldots, 12$, we have

$$a_{i1} + a_{i2} + a_{i3} + a_{i4} + a_{i5} + a_{i6} = \sum_{j=1}^{6} a_{ij} = a_{iT}.$$

In addition, for each $j = 1, \ldots, 6$, we have

$$a_{5,7-j} = a_{2j} + a_{6j}; \text{ moreover, } a_{5T} = a_{2T} + a_{6T}.$$

Example 2: AS149 is in three parts: Part I is an array a_{ij} where i = 1, 2 and j = 1, ..., 5; Part II is an array b_{ij} where i = 1, ..., 14 and j = 1, ..., 5; Part III is an array c_{ij} where i = 1, ..., 9 and j = 1, ..., 5. For all j, we have:

$$a_{1j} = b_{1j} + b_{2j} + b_{3j} + \ldots + b_{7j} = \sum_{i=1}^{7} b_{ij},$$

$$a_{2j} = b_{8j} + b_{9j} + b_{10j} + \ldots + b_{14j} = \sum_{i=8}^{14} b_{ij},$$

and, $$b_{ij} = c_{1j} + c_{2j} + c_{3j} + \ldots + c_{9j} = \sum_{i=1}^{9} c_{ij}.$$

Example 3: AS38 is an array a_{ij} where i = 1, ..., 6 and j = 1, ..., 18. Many of the pendants have first level subsidiaries s_{ijk}, where s_{ijk} is associated with a_{ij} and k = 1, ..., 11. On this quipu, we have:

$$a_{ij} = a_{2j} + a_{3j} = \sum_{i=2}^{3} a_{ij} \text{ for } j = 1, \ldots, 18;$$

$$a_{2j} = a_{4j} + a_{5j} + a_{6j} = \sum_{i=4}^{6} a_{ij} \text{ for } j = 1, \ldots, 18;$$

and, $$s_{2jk} = s_{3jk} + s_{4jk} + s_{5jk} + s_{6jk} = \sum_{i=3}^{6} s_{ijk}$$

for k = 3, 5, 6, ..., 11.

Another way in which sums play a role is where sets of numbers systematically have the same sum although the sum itself does not appear on the quipu. This would occur, for example, if there was an array a_{ij} where i = 1, ..., N and j = 1, ..., M such that

$$S_i = \sum_{j=1}^{M} a_{ij} \quad \text{or} \quad S_j = \sum_{i=1}^{N} a_{ij}$$

had the same value for all i or all j, respectively.

A small quipu of this type is AS141 which is an array a_{ij} where i = 1, 2 and j = 1, ..., 10. In this case, we have:

$$a_{11} + a_{12} + \ldots + a_{15} = a_{16} + a_{17} + \ldots + a_{1,10}$$
$$= a_{21} + a_{22} + \ldots + a_{25}$$
$$= a_{26} + a_{27} + \ldots + a_{2,10}.$$

In summation notation, one can write

$$\sum_{j=1}^{5} a_{1j} = \sum_{j=6}^{10} a_{1j} = \sum_{j=1}^{5} a_{2j} = \sum_{j=6}^{10} a_{2j}.$$

As a consequence, this also means that

$$\sum_{j=1}^{10} a_{1j} = \sum_{j=1}^{10} a_{2j}.$$

A larger quipu having this type of structure is AS136. This quipu is in two parts: Part I is an array a_{ij} where i = 1, ..., 10 and j = 1, ..., 9 and Part II is an array b_{ij} where i = 1, ..., 10 and j = 1, ..., 9. Here, using summation notation, we have:

$$\sum_{i=1}^{10} (a_{i4} + a_{i5}) = \sum_{i=1}^{10} (b_{i4} + b_{i5}) , \quad \sum_{j=1}^{9} a_{4j} = \sum_{j=1}^{9} b_{4j} ,$$

$$\sum_{j=2}^{9} a_{6j} = \sum_{j=2}^{9} b_{6j} , \quad \text{and} \quad a_{10,j} = b_{10,j} = 0 \quad \text{for}$$

j = 1, ..., 9.

Finally, a more complex quipu that combines both summation ideas is AS100. This quipu is in three parts: Part I is an array a_{1j} where j = 1, ..., 6; Part II is an array b_{ij} where i = 1, ..., 6 and j = 1, 10; and Part III is an array c_{ij} where j = 1, 2, 3. It also has a dangle end cord, d. On this quipu,

$$d = a_{11} + a_{12} + \ldots + a_{16} = \sum_{j=1}^{6} a_{1j} \, ,$$

$$a_{1k} = b_{k1} + b_{k2} + \ldots + b_{k,10} = \sum_{j=1}^{10} b_{kj} \text{ for } k = 1, \ldots, 6,$$

and $c_{11} = c_{12} = c_{13} = 0.$

Also, using summation notation for brevity, we have:

$$\sum_{i=1}^{6} b_{i3} = \sum_{i=1}^{6} b_{i7} \, , \quad \sum_{i=1}^{6} b_{i5} = \sum_{i=1}^{6} b_{i6} \, , \quad \text{and}$$

$$\sum_{i=1}^{6} b_{i8} = \sum_{i=1}^{6} b_{i,10}.$$

Finally, we also have $\displaystyle\sum_{i=1}^{6} b_{ik} = \sum_{j=1}^{10} b_{kj}$ for $k = 2$ or $4.$

There are subsidiaries on the quipu but we have not yet been able to identify any consistent relationships including them. Notice that on this quipu, the dangle end cord carries the sum of the values in Part I which also is the sum of all of the values in Part II. In general, a dangle end cord carries some number, such as this grand total, prominently associated with the quipu. It does not, however, necessarily carry a value related to summation.

Some quipus give the appearance of being "worksheets" in that they are tied together and similar values appear on them in different contexts. Nevertheless, quipus are not calculating devices, they are only records. The fact that a cord contains a value that is the sum of the values on other cords gives no information about the arithmetic process involved. In some cases, we believe it more likely that what we are identifying as the sum is a value that was decided first and then separated into parts. Where the parts bear a strikingly consistent relationship to the whole, it would seem to be evidence that the process of division was involved.

An example of such a quipu is AS120. This is an array a_{ij} where $i = 1, ..., 4$ and $j = 1, ..., 8$ with one subsidiary on each a_{i3}. For ease of notation, call the subsidiaries a_{is}. The values on the quipu range from 102 to 43,372. Here,

$$a_{2j} + a_{3j} + a_{4j} = a_{1j} \quad \text{for} \quad j = 1, ..., 8, s.$$

However, with a maximum error of 1.2 percent, $a_{2j} \div a_{1j} = .342$, $a_{3j} \div a_{1j} = .425$, and $a_{4j} \div a_{1j} = .235$, for $j = 2, ..., 8, s$. (If 796 is subtracted from a_{31} and added to a_{41}, these equations also hold for $j = 1$.)

The arithmetic relationships of values on several quipus are explored in greater detail in Ascher and Ascher (1980). Here, we close the discussion of arithmetic and statistical quipus by noting that our examination of values on a quipu and their organization can lead to hypotheses of intent and hypotheses of possibly related phenomena. For us, each quipu becomes a puzzle: one reminds us of a difference table; in another, the number of perfect squares is striking; on several, many values seem to be multiples of some single value as if dealing with a basic unit. These hypotheses, however, are difficult to substantiate and must, therefore, remain only hypotheses.

NON-NUMERICAL INFORMATION ON QUIPUS

Many quipus combine knotted cords that can be interpreted as magnitudes with cords that can be interpreted as number labels. Some others contain only number labels. The Spanish chronicles suggest many uses for quipus, some of which involve the transmission of non-numerical information. And, as the sole recording system used by the Incas, the quipus evidently were sufficient for their varied needs. When considering the possible uses of a system restricted to

digits variously assorted, one needs only remember that
digital computers store only strings of 1's and 0's. We
impute different meanings to different collections of them
and thereby store, process, and retrieve a variety of numeric
and non-numeric information.

On quipu cords where the knot clusters do not conform to
the standard base ten positional arrangement for single values or multiple values, the cords are interpreted as carrying
number labels. Then, corroboration of the interpretation is
sought in their consistency within the logical structure of
the quipu.

An example of a quipu carrying both magnitudes and number
labels is AS145. By spacing, it is in two parts: Part I is
an array b_{ij} where i = 1, ..., 4 and j = 1, ..., 7; Part II
is an array c_{ij} where i = 1, 2, 4 and j = 1, ..., 14 and
where for i = 3, j = 1, ..., 16. In Part I, the groups
alternate in color and also as to whether they are number
labels or magnitudes. In Part II, within each group, the
colors alternate. As in Part I, the data type alternates
with the color. Thus, b_{ij} is color 1 and a label when i is
odd; b_{ij} is color 2 and a magnitude when i is even. Also,
c_{ij} is color 1 or color 3 and a label when j is odd; c_{ij} is
color 4 and a magnitude when j is even.

On quipus where all the data are single digits with no
particular arithmetic relationships, magnitudes and number
labels are indistinguishable. In the cases where there are
the same few items of data appearing repeatedly, it seems
more plausible that they are number labels. An example of
such a quipu is AS97. It contains sixty-four pendants which
carry only single digits. One is a 2 and one is a 4 while

the other sixty-two are 0, 1, or 5. Another more complex
quipu (AS57) which also has this characteristic is particularly interesting because of its special construction features. Among these features are specially colored longer
cords with no knots and differently colored cord wrappings
around the main cord. They serve as additional ways of separating groups. The distinguishing markers combine with color
and spacing to form the quipu into two parts consisting of
seven and seventeen groups respectively. Each of these have
twenty pendant groups further separated into five subgroups
of four pendants each. Of the 340 pendants in the second
part, each carries a 0, 1, 2, or 4, except for one which
carries a 3.

TRANSFORMATIONS AND CORRESPONDENCES ON QUIPUS

Some quipus have been found physically linked together.
In several cases where they carry numerical information, the
values on them can be arithmetically related to each other.
In one case, where the data seems to be number labels, the
two connected quipus have a most intriguing relationship.
One of them has a persistent ABA pattern and the other has a
persistent AABA pattern, but, by color, the groups on one can
be matched to the groups on the other. The ABA pattern is on
a quipu having three groups of three cords each. In the
first and third groups, each pendant has one subsidiary,
while in the middle group each has two subsidiaries. Within
each group, the knot clusters on the first and third pendants
are the same but different for the middle pendant. Also,
within each group, the knot clusters on the subsidiaries of
the first and third pendants are the same but different for
those of the middle pendant. The second quipu, with an AABA
pattern, instead has four groups of four pendants each. In

the first, second, and fourth groups, each pendant has one subsidiary, while the third group differs and has no subsidiaries. Within each group, knot clusters are the same on the first, second, and fourth pendants but different on the third and, again, within each group, the knot clusters on the subsidiaries of the first, second, and fourth pendants are the same with the third's different. Thus, the elements creating the patterns are the same on both quipus. Moreover, the cord groups on the two quipus are in 1-1 correspondence in terms of color. The association is:

Quipu I	Quipu II
Pendant groups 1, 2, 3	- Pendant groups 1, 2, 4
Subsidiaries of groups 1, 3	- Subsidiaries of groups 1, 4
Higher subsidiaries of group 2	- Subsidiaries of group 2
Lower subsidiaries of group 2	- Pendant group 3

Similarly, there can sometimes be found on quipus which internally differ from part to part, an overall pattern with patterned relationships between the parts. An example is a quipu (AS98) which has fifteen groups, some with six pendants and some with five pendants and two subsidiaries on the first pendant. The color sequences of the different size groups distinctly differ from each other but involve the same five colors. Let Y be a group that has five pendants, two subsidiaries and a particular color pattern; X be a group of six pendants and a different particular color sequence; and a prime denote that a color sequence has been modified by substituting color 3 for colors 1 and 2. The entire quipu configuration, consisting of 100 pendants and subsidiary cords, can then be summarized as: XYYXYYXYYX'Y'Y'X'Y'Y'. As for the data on the quipu, it is numerical. The first pendant in each X (or X') group carries the sum of the ten values in the two following Y (or Y') groups. The rest of the pendants in each X (or X') group carry zero or are blank. (We hypothesize that this is an incomplete quipu and that these are blank cords which have not yet been knotted. We suspect that the X (or X') groups are sum groups and the five blank cords in each would cary the sums of the pairs of values in the corresponding positions in the two following Y or Y' groups.)

Another example of patterned relationships within an over-all pattern involves four quipus. Two of the quipus are in one museum and the other two are in different museums on two different continents. The fact that most quipus have little or no contextual information with them means that these quipus may or may not be related insofar as specific place of origin. Or, they might or might not be different representa-tives of some general form that was in use. Each of the four quipus have groups of nine pendants with a similar color sequence. Let A, B, D represent different solid colors; M represent any color mixture; and X be the nine color sequence AAMMAAAAA. A replaced by B will be denoted by ', A replaced by D by *, and M replaced by D by ¯. The four quipus can then be described as:

AS197: X
AS38:   XX̄*XX*XX*XX*XX*XX*
AS136: XXXXXXXXXXX̄'X̄'X̄'X̄'X̄'X̄*X̄*X̄*X̄*X̄*
AS140: X̄X̄X̄X̄X̄X̄X̄X̄X̄X̄X̄X̄X̄X̄X̄X̄X̄X̄.

These all carry numerical information. Calling the arrays a_{ij} where $i = 1, \ldots, 9$ and $j = 1, \ldots, N$ (N differs for each quipu), a slight numerical consistency can be found. Namely, for all j,

$$a_{1j} > a_{3j} > a_{2j} \; ; \; a_{8j} > a_{9j} \; ; \; a_{1j} = \max_i a_{ij}.$$

A final example is a quipu (AS15) which, at first glance, looks peculiar and uninteresting. It is peculiar in that it

has a long "tail" (a cord attached to the main cord differently than are other cords), which has on it single knots
interspersed with very short cord attachments that we call
"flags". The rest of the quipu is twelve pendant cords: the
first nine are either color C1 or C2 with one single knot
each and the last three are three different colors, all with
no knots. When the tail is carefully examined, the flags are
seen to be of color C1 or C2 forming a color sequence C1, C2,
C1, C1, C2, C2, C1, C2, C2. The knots on the tail are
separated by the flags into groups of 3, 1, 2, 1, 2 knots
while the flags are separated by the knots into groups of 1,
2, 1, 3, 2 flags. Thus, there are nine pendants with single
knots, nine single knots on the tail, and nine flags on the
tail. The color sequence of the nine knotted pendants is
exactly the same as the color sequence of the nine flags on
the tail. The flags are separated into five groups and the
tail knots are separated into five groups. The sizes of the
groupings correspond to each other: for i = 1, ..., 5,
size of ith knot group = size of $(5-i)^{th}$ flag group.

A CLOSING THOUGHT

We do not know whether the quipumakers were recording
information that was ordered and patterned or whether they
were imposing order and pattern on it as they recorded. But
then, there is an endless argument as to whether mathematicians discover order and patterns that already exist or
impose the order and invent the patterns in the process of
exhibiting them. In either case, the mathematical ideas
embodied in the quipus are the fundamental ideas of number,
spatial configuration, and logic. Quipus are mathematical
records in that they contain numerical data but, over and
above that, they are an expressive symbolic system. Two

essays in the recently published Mathematics Today use the same quote by Whitehead to capture a most significant aspect of mathematics. That same quote is fitting here:
"The notion of the importance of pattern is as old as civilization. Every art is founded on the study of pattern. ... Thus the infusion of patterns into natural occurrences and the stability of such patterns, and the modification of such patterns is the necessary condition for the realization of Good. Mathematics is the most powerful technique for the understanding of pattern, and the analysis of the relation of patterns."[4]

NOTES

1. The Wenner-Gren Foundation for Anthropological Research provided partial financial support for our work.

2. All photographs were taken by the author and her husband Robert Ascher. The quipus in figures 10.5, 10.6, and 10.7, respectively, are in the collections of P. Dauelsberg, Arica, Chile; Peabody Museum, Cambridge, Mass.; and Museo Nacional de Anthropología y Arqueología, Lima, Peru.

3. Identification of actual quipus are by tags made up of one or two letters and a number. All published quipus are identified by this system in Ascher and Ascher (1978). The letters refer to the authors and the numbers to the individual quipus in chronological order of publication by author. The detailed descriptions of all quipus used as examples in this article are published in the above source. In order to avoid the inclusion of extraneous details, some of the examples here use very slightly streamlined descriptions of the actual data. In no case, however, does this affect the point being made.

4. The quote appears in the essays "Mathematics -- Our

Invisible Culture," Allen L. Hammond, p.31, and "The Relevance of Mathematics," Felix E. Browder and Saunders MacLane,
p.348, in Mathematics Today, Lynn A. Steen, editor, Springer-
Verlag, New York, 1978.

11. The Mathematical Notation of the Ancient Maya

Michael P. Closs

INTRODUCTION

The most sophisticated development of mathematics indigenous to the New World occurred among the ancient Maya who
inhabited a region encompassing Guatemala, Belize, the western parts of Honduras and El Salvador, and the lowlands of
southern Mexico (the states of Yucatan, Campeche, Quintana
Roo, most of Tabasco and the eastern part of Chiapas). These
lands are still occupied by the descendents of the ancient
Maya, more than two million of whom still speak Mayan languages today. The region also contains the numerous ruins of
ancient Mayan settlements from which have come thousands of
carved stone monuments and ceramic vessels containing hieroglyphic texts. In addition, three Maya books, called
codices, are extant. Sadly, the ability to read the Maya
script has been lost as has so much of ancient Maya history.
However, there has been real progress in the painstaking task
of deciphering the Maya hieroglyphs, usually referred to as
simply "glyphs". This effort, begun almost a hundred years
ago, is continuing today, with much more optimism than was
the case only a few decades ago. Historians of mathematics
are in a relatively good position to make use of the results
of this glyphic research because most of the calendrical and
chronological records of the Maya are well understood and it
is within that context that Maya mathematics must be studied.

Books on the history of mathematics do not always refer to
Mayan developments and when they do so it is almost always in
a cursory manner. For example, it may be mentioned that the

Maya utilized a system of positonal notation which incorporated a zero and perhaps a brief description of Maya bar and
dot numeration may be attempted. However, many other incred-
ibly attractive Maya mathematical notations are never discussed and the contexts and purposes of Maya mathematics are
neglected. Unfortunately, one does not even find references
as to where this type of information is available. In fact,
all too often, one encounters the unjustified notion that
more detailed information on the mathematical development of
the Maya is not possible because of the ravages of the Spanish conquest and the mystery of an undeciphered script.

The intent of the present paper is to provide a concise
survey of Maya mathematical symbolism and to illustrate some
of the contexts in which it is employed.[1] In order to reach
this goal it is first necessary to briefly examine the Maya
calendar and the Maya system of chronological reckoning.
Only those concepts and structures needed for the specific
purposes of this paper will be considered.

THE MAYA NUMBER SEQUENCE

In the study of Maya hieroglyphic writing it is customary
to use Yucatec terms for number words and calendar names.
These are transcribed according to colonial Yucatec orthography. Briefly, the letter x is for š, prevocalic u is for
w, c is for k, k is for the midvelar glottalized stop k', and
doubled vowels (except with syllable-initial u for w) are for
vowels interrupted or checked by a glottal stop. Despite the
use of Yucatec terms, it should be remembered that there are
some thirty different Mayan languages and Yucatec is not the
only, nor necessarily the most extensive, idiom underlying
the glyphic inscriptions.

The Yucatec number words reveal a clear vigesimal

structure with a decimal stratum evident in the numeration
from thirteen through nineteen.

1	hun	11	buluc
2	caa, ca	12	lahca
3	ox	13	oxlahun
4	can	14	canlahun
5	hoo, ho	15	hoolahun, hoolhun
6	uac	16	uaclahun
7	uuc	17	uuclahun
8	uaxac	18	uaxaclahun
9	bolon	19	bolonlahun
10	lahun	20	hun kal

Words for twenty, or score, in the Mayan languages are
kal, may, and uinic, or forms cognate to one or another of
these. The last is the term for "man" or "human being" and
in this context refers to the totality of his digits. The
other terms are apparently related to words for tying and
bundling and may reflect practices of counting and packaging
in ancient commerce and rendering of tribute.

Multiples of 20 follow a regular pattern up to 380, after
which comes hun bak, 'one 400'. For example, 40 is given by
ca kal, '2 score', 60 by ox kal, '3 score', and 380 by
bolonlahun kal, '19 score'.

The first six powers of twenty are given by unit terms as
follows:

20^1	kal	20^4	calab
20^2	bak	20^5	kinchil
20^3	pic	20^6	alau

The Cakchiquel equivalent of Yucatec pic, '8000', is
chuwi, which is also a word for "sack". Its use as a numeral

is said to derive from the custom of packaging cacao beans
-- an important commodity and also a medium of exchange -- in
quantities of 8000 to the bag.

Multiples of the higher powers of twenty are enumerated in
the same way as those of the first power.

There are two different methods of naming numbers that
intervene between the multiples of any power of twenty. In
the first system, prevalent in many Mayan languages today,
the intervening quantity was named and placed in the ordinal-
numbered score or other power of twenty to which it belonged.
The second method of expressing compound numerals was to use
a conjunction as we do, either expressed (<u>catac</u> in Yucatec)
or implied by juxtaposition of two orders of components, and
proceeding from the higher-order to the lower-order components. Thus, for example, 51 could be either <u>buluc tu yox
kal</u>, 'eleven in the third score', or <u>ca kal catac buluc</u>, 'two
score and eleven'.

The first system is complicated by three types of ellipsis
which enter into it (Lounsbury 1978, pp.762-763). First,
between higher powers of 20, simple numerals represented multiples of the next lower power rather than multiples of unity
as might be expected. Secondly, the word for 2 was omitted
in reference to the second 20, the second 400, and so forth.
Finally, the syllable <u>tu</u> was dropped from those terms in
which it would be preceded by the words for 10 or 15. The
last two ellipses were complementary, that is, one numeral
expression could not contain both. However, either of them
could be combined with an ellipsis of the first type.

THE CALENDAR ROUND

The most important calendrical cycle of the Maya, as of
other Mesoamerican peoples, was a ritual calendar of 260

days. In Mayanist literature it is designated as the tzol-
kin, 'sequence of days', the sacred almanac, or the Sacred
Round. It is the product of a cycle of 13 day numbers with a
cycle of 20 day names. Although the day names varied in different Mayan languages, they were represented by the same
glyphs, including variants, wherever Maya inscriptions are
found. The sequence of day names (expressed in Yucatec, as
is the custom) and their corresponding glyphs are shown in
Table 11.1. The first two columns of glyphs are typical of
the inscriptions and the last two are typical of the codices.
The glyphs in the first and third columns are of a type often
referred to as "symbolic variants", while those in the second
and fourth are referred to as "head variants". The day
glyphs in the inscriptions are characterized by being placed
within cartouches which often rest on trinal supports. The
days of the Sacred Round begin with 1 Imix, 2 Ik, 3 Akbal,
and continue in this fashion up to the thirteenth day, 13
Ben. The next day, the fourteenth in the calendar, is 1 Ix
[note that 14 ≡ 1, mod 13], the one after that is 2 Men
[15 ≡ 2, mod 13], and so on. After the twentieth day, 7 Ahau
[20 ≡ 7, mod 13], the day names begin to repeat and one
arrives at the twenty-first day 8 Imix [21 ≡ 8, mod 13], the
twenty-second day 9 Ik [22 ≡ 9, mod 13], and so on. This
pattern continues until 260 days have elapsed, one complete
cycle of the Sacred Round.

   The Maya also employed a 365-day calendar often referred
to as the Vague Year because it is a whole day approximation
to the sidereal year and does not preserve an alignment with
the seasons over long periods of time. It was made up of 18
named months of 20 days each and a residual period of 5 days.
The names of the months and the residue vary considerably
from one Mayan language to the next but their glyphs,

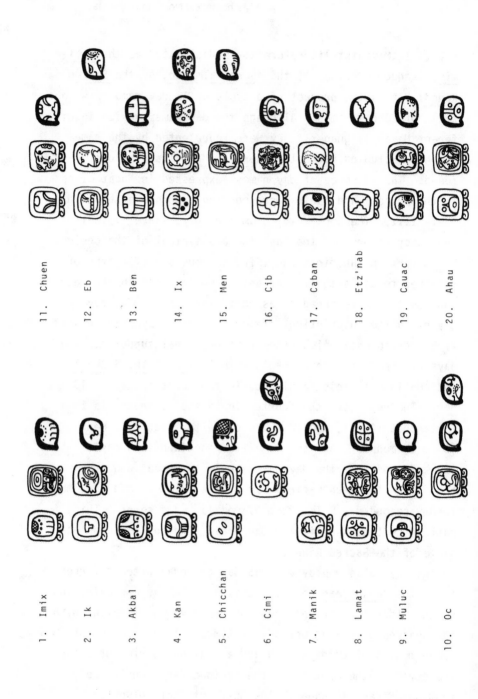

11. Chuen

12. Eb

13. Ben

14. Ix

15. Men

16. Cib

17. Caban

18. Etz'nab

19. Cauac

20. Ahau

1. Imix

2. Ik

3. Akbal

4. Kan

5. Chicchan

6. Cimi

7. Manik

8. Lamat

9. Muluc

10. Oc

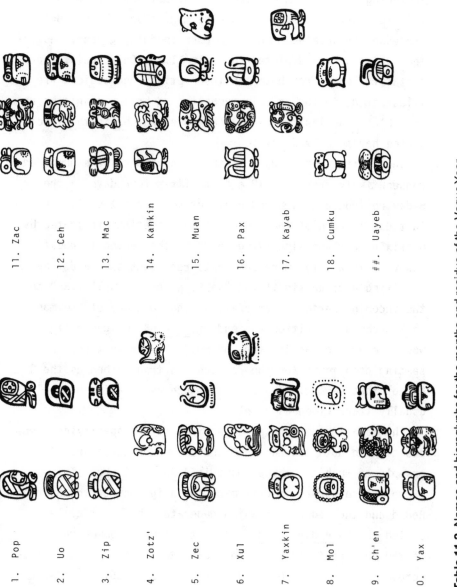

11. Zac

12. Ceh

13. Mac

14. Kankin

15. Muan

16. Pax

17. Kayab

18. Cumku

##. Uayeb

1. Pop

2. Uo

3. Zip

4. Zotz'

5. Zec

6. Xul

7. Yaxkin

8. Mol

9. Ch'en

10. Yax

Table 11.2. Names and hieroglyphs for the months and residue of the Vague Year.

including variants, are common throughout the area of the inscriptions. The sequence of the months and the residue, expressed in Yucatec, and their corresponding glyphs, including variants, are depicted in Table 11.2. Once again, representative forms from both the inscriptions and codices are illustrated, including both symbolic and head variants where possible. The last entry in the table, Uayeb, is the special period having only 5 days.

The first nineteen days in the 20-day months are numbered consecutively from 1 to 19 and the first four days in the 5-day residue are numbered consecutively from 1 to 4. The last day in each of these periods was sometimes indicated by prefixing a glyph signifying "end" to the period in question. However, it was far more common to represent such a day by prefixing a glyph signifying "seating" or "installation" to the incoming period. For example, the 20th day of the month Yaxkin could be written as "end of Yaxkin" but generally it would be written as "seating of Mol". By convention, the seating of a month (or the residue) is transcribed as the 0 day of that period. Thus, a date such as "seating of Mol" is usually transcribed as 0 Mol.

The Maya liked to describe a given day by specifying its position in both the Sacred Round and the Vague Year. An example of such a combined date is 4 Ahau 8 Cumku. This example also serves to determine the alignment of the Sacred Round and the Vague Year which generates the joint cycle called the Calendar Round. Since the lowest common multiple of 260 and 365 is 18,980, the Sacred Round and the Vague Year combine to form 18,980 (= 73 × 260 = 52 × 365) Calendar Round dates.

BAR AND DOT NUMERALS

The common numerals of the Maya were composed of bars with value 5 and dots with value 1. Numbers from 1 to 19 were represented by an economical combination of bars and dots yielding the desired value. Numbers could be written horizontally or vertically. If the bars were horizontal the dots were placed above them; if the bars were vertical the dots were placed to their left. Frequently, non-numerical crescents or other fillers were used to achieve a more esthetic balance for the numeral.

Bar and dot numerals were heavily employed in recording Calendar Round dates. Indeed, it will be recalled that the numbers 1 to 13 appear as day coefficients in the Sacred Round. In addition, the numbers 1 to 19 appear as coefficients in the first 19 days of the 20-day months of the Vague Year and 1 to 4 appear in the first 4 days of the 5-day residual period. Several examples of Calendar Round dates in which the numerical coefficients are expressed by bar and dot numerals are shown in figure 11.1, a-d. As noted earlier, the last day of a month or of the residual period could be represented by a glyph signifying "end" prefixed to the corresponding period. When, as was usually the case, the last day was expressed as the seating day of the incoming period a glyph signifying "seating" was prefixed to that period. For the sake of completeness, examples of dates employing these special glyphs are also illustrated in figure 11.1, e-h. The main sign in the prefix of the month glyphs in figure 11.1, e-f, is a tun glyph which in this context signifies "end" or "final". The same glyph is used to designate a 360-day period called tun in other contexts to be considered later.

It is natural to expect that numbers were also used to

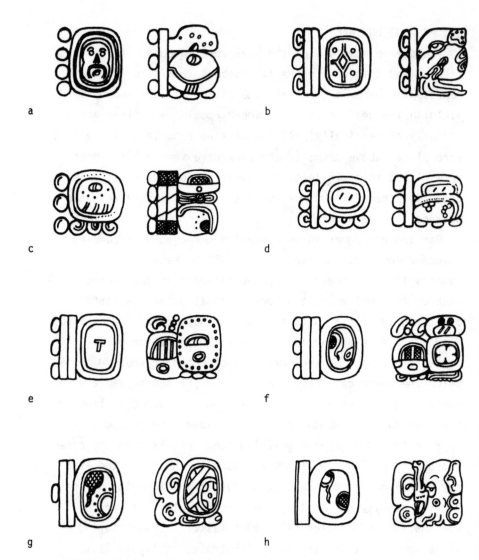

Fig. 11.1. Calendar Round dates: (a) 4 Ahau 8 Cumku, Palenque, Tablet of the Cross, D3-C4; (b) 11 Lamat 6 Xul, Palenque, Tablet of the Cross, P10-Q10; (c) 3 Imix 14 Ch'en, Yaxchilan, Lintel 25, 01-N2; (d) 6 Chicchan 8 Yax, Yaxchilan, Lintel 27, A1-B1; (e) 13 Ik end of Mol, Palenque, Tablet of the Cross, C9-D9; (f) 13 Caban end of Yaxkin, Aquateca, Stela 1, B11-A12; (g) 11 Caban 0 Pop, Palenque, Tablet of the Cross, Q2-P3; (h) 5 Caban 0 Zotz', Palenque, Tablet of the Cross, S12-R13.

enumerate objects of trade and tribute. However, unlike the situation for the Aztec, no Maya records of this type have survived. Nevertheless, there are some instances of a similar usage of number in which offerings are enumerated. Probably the clearest example of such an enumeration is that found in the cycle of New Year's ceremonies prescribed in the Dresden Codex. Here, offerings of a variable number of nodules of copal incense, pom, are specified for each of the ceremonies (Lounsbury 1973, pp.105-111). Three of these enumerations are depicted in figure 11.2.

Another, somewhat unusual, usage of number is found in the names of many gods and goddesses of the Maya. The name glyphs of two such deities, Bolon Yocte, 'Nine Strides', and Lahun Chan, 'Ten Sky', have been identified and are reproduced in figure 11.3, a-b. There are other gods in the codices whose name glyphs carry numerical prefixes but whose Maya names remain unknown. For example, in the Madrid Codex there is a god of death, known as God Q, whose name glyph includes the numeral 10. Another god, this time a benevolent deity, known as God R, appears in the Dresden Codex with a coefficient of 11. The name glyphs of God Q and God R are depicted in figure 11.3, c-d. The Dresden Codex also contains name glyphs for the moan bird, believed to symbolize the 13 heavens of Maya cosmology and to be a deity of clouds and rain. Its name has two variant forms, shown in figure 11.4, e-f, the first being a moan bird head with a prefix of 13 and the second being the sky sign with a prefix of 13.

THE MAYA CHRONOLOGICAL COUNT AND A GLYPH FOR ZERO

The Maya measured time intervals between calendar dates by a composite chronological count consisting of a vigesimal

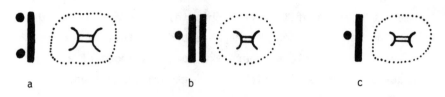

Fig. 11.2. Offerings of nodules of copal incense: (a) 7 nodules of copal, Dresden 26; (b) 11 nodules of copal, Dresden 27; (c) 6 nodules of copal, Dresden 28.

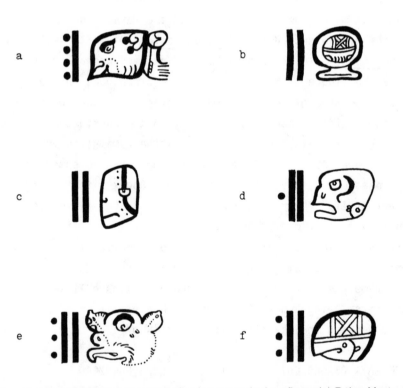

Fig. 11.3. Some Maya deity names having numerical prefixes: (a) Bolon Yocte, Dresden 60b; (b) Lahun Chan, Dresden 47b; (c) God Q, Madrid 86b; (d) God R, Dresden 5b; (e) Moan bird, Dresden 8b; (f) Moan bird, Dresden 7c.

count of tuns (360-day periods) and distinct counts of
uinals (20-day periods) and kins (days) (Closs 1977b). It
has been shown that in some contexts, such as that described
here, the 360-day period may have been referred to as the
haab' (Fox and Justeson 1984, pp.52-53). Thus, in similar
contexts discussed in the remainder of the paper, the glyph
read as tun, may in fact have been read as haab'. The
respective units of these three base counts are related by
the reduction formulas:

$$20 \text{ kins} = 1 \text{ uinal}$$
$$18 \text{ uinals} = 1 \text{ tun.}$$

In addition to glyphs for the three chronological units,
the Maya had glyphs for 20, 20^2, 20^3, 20^4, and 20^5 tuns.
These larger periods are the first five orders or levels in
the vigesimal tun count and are respectively called katuns,
baktuns, pictuns, calabtuns, and kinchiltuns. It should be
noted that the terms kin, uinal, tun, and katun are well
attested in the post-conquest documentary sources. However,
the terms for the four larger periods are modern contrivances
formed by prefixing the Yucatec terms for 400, 8000, 160,000,
and 3,200,000 to the tun. Hieroglyphs for all of the above
periods are illustrated in figures 11.4 and 11.5.

Of the various glyphs in figure 11.4, only the first two
are known to linguistically represent kin, 'day, sun'. In
fact, the second of these depicts the head of the old sun
god. Although the other glyphs replace the kin they are not
necessarily linguistically identical to it and may represent
other semantic notions such as "night" or "darkness". The
glyph in figure 11.4d is a monkey head and that in figure
11.4e is a shell. Linda Schele (1977, pp.52-53) has shown
that the glyph in figure 11.4f is a brocket deer head. The
one in figure 11.4k shows a kin glyph nestled between glyphs

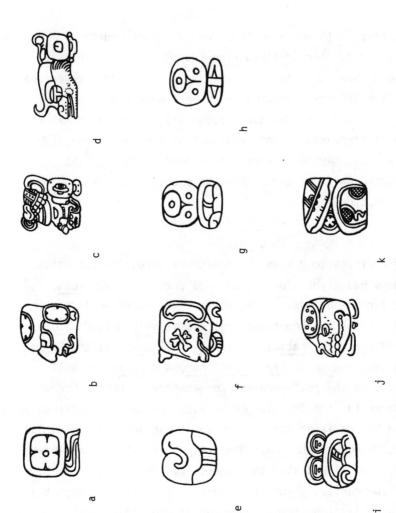

Fig. 11.4. Hieroglyphs for kin and its substitutes.

a

b

c

d

e

f

g

h

i

j

k

Fig. 11.5. Hieroglyphs for the uinal, tun and higher chronological periods.

signifying "sky" and "earth". It has been suggested that
this combination represents the sun at the horizon and hence
signifies "dawn".

It was normal procedure in the monumental inscriptions to
record a chronological count by an ascending or descending
series of period glyphs with appropriate coefficients. It is
worth noting that the kin variants in figure 11.4, b-d, are
restricted to chronological counts recorded by descending
series while the ones in figure 11.4, e-j, are restricted to
counts recorded by ascending series. The kin glyph in figure
11.4a and the rare variant in figure 11.4k, always occurring
with a coefficient of 1, are exceptional in that they appear
with both types of series .[2] The glyphs for the uinal and
higher periods, shown in figure 11.5, do not exhibit a similar restricted usage. In transcribing a chronological count,
it is customary to write the number of its kins, uinals,
tuns, etc., in decreasing order of the periods, and to separate the respective numbers by points. Thus, for example, a
count of 17 kins, 11 uinals, 2 tuns, and 13 katuns is transcribed as 13.2.11.17.

A chronological count which links two recorded calendar
dates is commonly described as a "Distance Number". The same
term applies to a chronological count which links a recorded
date to an earlier or later event, glyphically specified,
although not labeled by a date. In all but a handful of
cases Distance Numbers in the inscriptions are represented by
an ascending series of period glyphs. It is also interesting
to note that frequently the kin glyph in a Distance Number is
omitted and its coefficient is prefixed to the count of the
uinals.

It is apparent that if a chronological count lacks one or
more intermediate periods then a problem arises in recording

it. Either the glyphs of the missing periods must be deleted
or they must be written with an explanatory prefix signalling
that the corresponding periods have a zero count. In all but
a few exceptional cases the Maya chose the latter alternative. Several Distance Numbers, some exhibiting the use of
the zero glyph, are reproduced in figure 11.6. The normal
glyph used in the inscriptions to indicate a zero count is
depicted in figure 11.6b prefixed to the <u>uinal</u> glyph, and in
figure 11.6c prefixed to both the <u>uinal</u> glyph and the <u>kin</u>-
substitute.[3] The reading order for the glyphs in figure 11.6,
a-d,f is the reverse of that occurring in the transcribed
Distance Numbers while for the glyphs in figure 11.6e it is
the same as in the transcribed Distance Number. The order of
reading glyphs in Maya texts with more than one column has a
fixed pattern: one proceeds in column pairs, from left to
right and top to bottom. The Distance Number in figure 11.6e
is a rare example expressed by a descending series. The
remaining Distance Numbers in figure 11.6 are of normal type
and are expressed by ascending series.

COMPUTATIONAL TECHNIQUES

There are two common calendrical problems often encountered when studying the Maya inscriptions. These were also
problems which must have been faced by the Maya scribes. The
first is to determine the Calendar Round date reached when
one counts through a given interval of time, forwards or
backwards, from a prescribed Calendar Round date. The second
and more difficult problem is to determine the minimum chronological interval separating two given Calendar Round dates.
Of the various solutions to these problems which have been
proposed, the ones outlined here are probably closest in
spirit to the techniques used by the ancient Maya, although

a

b

c

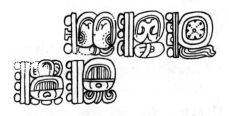

d

e

f

Fig. 11.6. Distance Numbers from the inscriptions: (a) 4.19, Piedras Negras, Stela 3, F6; (b) 3.0.17, Kuná-Lacanhá, Lintel 1, K1-L1; (c) 7.0.0, Yaxchilan, Lintel 31, K3-L3a; (d) 4.1.10.18, Palenque, Temple of the Inscriptions, West Panel, T6-S7; (e) 7.14.14.11.12, Palenque, Temple of the Foliated Cross, Alfardas, D1-F1; (f) 9.12.18.5.16, Palenque, Tablet of the Sun, C14-C16.

the form used to express them is entirely modern.[4]

A Calendar Round date can be specified by a triple (t,v,y) where t, v, and y represent the positions of the given day in the trecena (the cycle of 13 day coefficients in the Sacred Round), the veintena (the cycle of 20 day names in the Sacred Round), and the Vague Year, respectively. For example, the date 4 Ahau 8 Cumku has trecena position 4, veintena position 20, and Vague Year position 348. The last datum is obtained by noting that 8 Cumku is the 8th day of the 18th month and so is $(17 \times 20) + 8 = 348$ days after the seating of the Vague Year on 0 Pop.

Since most of the chronological computations in the Maya inscriptions span an interval of less than 13 baktuns, the algorithm presented for the first problem will only deal with this situation. The technique can be expanded to cover other larger intervals if required. Let (t,v,y) be the date which is reached after a chronological count of n_5 baktuns, n_4 katuns, n_3 tuns, n_2 uinals, and n_1 kins forward from a date (t_0,v_0,y_0). Then t, v, and y are given by the following:

$$t = t_0 - n_5 - 2n_4 - 4n_3 + 7n_2 + n_1, \text{ mod } 13$$

$$v = v_0 + n_1, \text{ mod } 20$$

$$y = y_0 + 190n_5 - 100n_4 - 5n_3 + 20n_2 + n_1, \text{ mod } 365.$$

If the count is backwards from a given date, say (t,v,y), the above equations can be solved for t_0, v_0, and y_0 to obtain the earlier date (t_0, v_0, y_0).

The above algorithm can be used to verify many calendrical and chronological statements in the inscriptions. For example, consider the text in figure 11.7 dealing with events in the life of a ruler of Yaxchilan known as "Bird-Jaguar". The text may be paraphrased as follows:[5]

A - B :   4   ...

G - H :   1

       2

       3

       4

       5

I - J :   1

       2

Fig. 11.7. A hieroglyphic inscription dealing with the reign of Bird-Jaguar of Yaxchilan, Lentel 29, A4, Lintel 30, G1-H5, and Lintel 31, I1-I2 (redrawn from Graham and Von Euw 1977-1979).

A4, G1   [On] 8 Oc 13 Yax
H1   was born
G2   Bird-Jaguar
H2   [title of Bird-Jaguar]
G3   "Lord of Yaxchilan".
H3a   [Its change was] 10 [kins], 5 uinals,
H3b   3 tuns, and
G4   2 katuns;
H4a   "then occurred"
H4b-G5   11 Ahau 8 Zec
H5a   when was seated
H5b   in the rulership
I1   Bird-Jaguar
J1a   [title of Bird-Jaguar]
J1b   3 katun lord
I2   "Lord of Yaxchilan".

The earlier part of the inscription preceding G1 has been
omitted except for the Sacred Round date at A4. This deleted
material contains chronological and astronomical data associated with the opening date 8 Oc 13 Yax. It also includes a
clause which is peculiar to a cycle of 819 days and a Distance Number linking it to the opening date. The text in
figure 11.7 consists of two sentences as indicated, the first
recording the birth of Bird-Jaguar and the second his inauguration as ruler of Yaxchilan. The corresponding dates for
these events are separated by the Distance Number at H3-G4.
The "3 katun lord" at J1b concerns the personal chronology
of Bird-Jaguar and signifies that he was in his 3rd katun of
life when this monument was erected.
     To verify the calculation in figure 11.7, one first notes
that the opening date 8 Oc 13 Yax corresponds to the triple

R   S

13

14

15

16

17

Fig. 11.8. An episode in the life of Kan-Xul of Palenque; Palenque, Tablet of the Cross, S13-S17 (after Maudslay 1889-1902).

(8,10,193). Next, one counts forward 2.3.5.10 to reach the
date (t,v,y) where

$t = 8 - 2(2) - 4(3) + 7(5) + 10 = 11$, mod 13

$v = 10 + 10 = 20$, mod 20

$y = 193 - 100(2) - 5(3) + 20(5) + 10 = 88$, mod 365.

Hence, the terminal date of the calculation is (11, 20, 88),
that is, 11 Ahau 8 Zec, precisely as recorded in H4b-G5.

In addition to checking calculations, the algorithm may be
used to recover information which has not been recorded in a
text. For instance, consider the text in figure 11.8, taken
from the Tablet of the Cross at Palenque. It can be read as
follows:[6]

S13	[Its change was] 16 [kins], 6 uinals,
R14	19 tuns, and
S14	1 katun
R15	[from the] birth [of]
S15	Kan-Xul
R16-S16	to (his) "accession in office"
R17-S17	[on] 5 Kan 12 Kayab.

The text recalls the birth and the inauguration of Kan-Xul
as ruler. Although abbreviated, it is similar in substance
to the previous inscription. In the present case, a separate
birth clause is missing but the birth date may be recovered
by counting backward from the accession date 5 Kan 12 Kayab,
or (5,4,332), through the Distance Number 1.19.6.16. The
birth date is then given by (t_0, v_0, y_0) where:

$t_0 = t + n_5 + 2n_4 + 4n_3 - 7n_2 - n_1$

$\quad = 5 + 2(1) + 4(19) - 7(6) - 16 = 12$, mod 13

$v_0 = v - n_1$

1 2 3

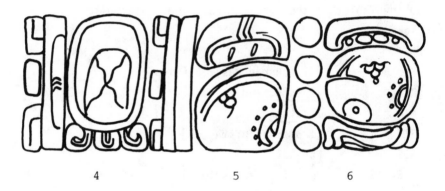

4 5 6

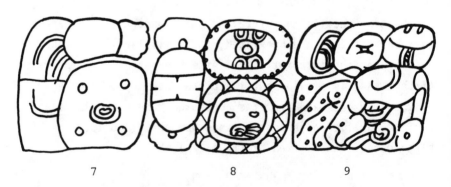

7 8 9

Fig. 11.9. The beginning of the mortuary text of Pacal of Palenque; Palenque, Temple of the Inscriptions, Sarcophagus Lid, 1-9 (after a drawing of Merle Greene Robertson in Lounsbury 1974).

$$= 4 - 16 = 8, \text{ mod } 20$$

$$y_0 = y - 190n_5 + 100n_4 + 5n_3 - 20n_2 - n_1$$

$$= 332 + 100(1) + 5(19) - 20(6) - 16 = 26, \text{ mod } 365.$$

Thus, Kan-Xul was born on (12,8,26), that is, 12 Lamat 6 Uo.

The second type of calendrical problem was to find the minimum chronological interval separating two Calendar Round dates, say (t_1,v_1,y_1) and (t_2,v_2,y_2). Let $\Delta t = t_2 - t_1$, mod 13, $\Delta v = v_2 - v_1$, mod 20, and $\Delta y = y_2 - y_1$, mod 365. Denote the minimal interval in the Sacred Round by ΔSR, the number of whole Vague Years contained in the Calendar Round interval by n_y, and the Calendar Round interval itself by ΔCR. Then

$$\Delta SR = 40(\Delta t) - 39(\Delta v), \text{ mod } 260,$$

$$n_y = \Delta SR - \Delta y, \text{ mod } 52,$$

and, $\Delta CR = 365(n_y) + \Delta y$, mod 18,980.

In order to illustrate the use of this algorithm consider the inscription in figure 11.9. This is the beginning of a text from the lid edge of a magnificent sarcophagus found in a vaulted tomb buried deep within the Temple of the Inscriptions at Palenque.[7] Pacal, perhaps the greatest of all Maya kings, was interred in the sarcophagus and the inscription opens with a declaration of his birth and death. The glyphs may be translated as follows:

1-2	[On] 8 Ahau 13 Pop
3	[he] was born;
4-5	[On] 6 Etz'nab 11 Yax
6	"4 times the tuns were seated"
7	[he] died;
8	K'ina Pacal
9	"Lord of Palenque".

The term K'ina in Glyph 8 is a title of rulership often

carried by the kings of Palenque. The preceding algorithm
can be used to calculate the interval between Pacal's birth
and death, modulo 18,980 days (or 52 Vague Years). The possibility of ambiguity caused by the periodic nature of the
solution is removed by Glyph 6. The reference to "seatings
of tuns" signifies the beginnings of katuns. Since four
katuns were begun during the lifetime of Pacal, it is clear
that his age at death is not less than 60 nor more than 100
years.

Now, 8 Ahau 13 Pop corresponds to (8,20,13) and 6 Etz'nab
11 Yax corresponds to (6,18,191). Thus, the algorithm yields
the following results:

$$\Delta t = 6 - 8 = -2, \bmod 13$$
$$\Delta v = 18 - 20 = -2, \bmod 20$$
$$\Delta y = 191 - 13 = 178, \bmod 365$$
$$\Delta SR = 40(-2) - 39(-2) = -2, \bmod 260$$
$$n_y = -2 - 178 = 28, \bmod 52$$
$$\Delta CR = 365(28) + 178 = 10,398, \bmod 18,980.$$

Because Pacal's life span was between 60 and 100 years, a
Calendar Round interval must be added to ΔCR to obtain his
age at death. To express ΔCR in the form of a Maya chronological count one rewrites it in the form

$$\Delta CR = 7200 + 8(360) + 15(20) + 18 = 1.8.15.18.$$

Similarly, the Calendar Round interval can be expressed in
Maya form by writing:

$$18,980 = 2(7200) + 12(360) + 13(20) = 2.12.13.0.$$

Thus, Pacal's age at death is given by

$$1.8.15.18 + 2.12.13.0 = 4.1.10.18$$

or, approximately 81 years.

It is interesting to note that the date 6 Etz'nab 11 Yax
is also recorded on the west panel of the Temple of the
Inscriptions at Palenque. This is followed by a glyph

signifying death, another giving Pacal's name, and a chronological count recording his age at death, namely 4.1.10.18.
This record has been illustrated in figure 11.6d.

THE LONG COUNT

Maya dates are frequently anchored in a remarkable system
of absolute chronology known as the "Long Count". This is a
framework used by the Maya to fix the chronological position
of all dates with respect to a base date far in the past.
That base date has Calendar Round position 4 Ahau 8 Cumku and
is represented by the formula 13.0.0.0.0 in the Long Count
chronology. The Long Count position of any date following
13.0.0.0.0, 4 Ahau 8 Cumku, is precisely the chronological
count linking the two dates. Thus, for example, the day following the base date has Calendar Round position 5 Imix 9
Cumku and Long Count position 1. Most of the stone monuments
bearing hieroglyphic inscriptions were erected more than 3000
years after the chronological base date and have Long Count
positions falling between 8.12.0.0.0 and 10.4.0.0.0. According to the most favored correlation of the Maya and Christian
chronologies this historic era runs from around A.D. 278 to
A.D. 909.

In many cases the Maya explicitly recorded the Long Count
position of the first date of an inscription. The sequence
of glyphs forming this chronological statement is known as an
"Initial Series" because of its usual position at the beginning of a text. In the inscriptions, but not in the codices,
Initial Series commence with a distinctive introductory glyph
often double or quadruple the size of other glyphs in the
text. This introductory glyph contains a variable central
element which is dependent on the month (or residual period)
of the Vague Year in which the date marked by the Initial

A   B

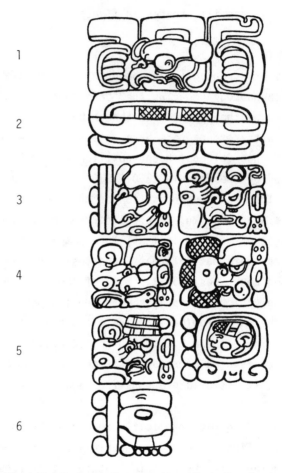

Fig. 11.10. The Initial Series on the east side of Stela C at Quirigua (after Maudslay 1889-1902).

Series falls. It is also an inviolable rule that the period
glyphs of an Initial Series be listed in decreasing order,
the reverse of the usual order for Distance Numbers. As a
result, the kin variants employed in Initial Series are of
the types exhibited in figure 11.4, a-d.

One of the chronological formulas used by the Maya to represent the base date of the Long Count is depicted on Stela C
at Quirigua and is illustrated in figure 11.10. The formula
is given in Initial Series form and begins with an introductory glyph occupying four glyph blocks. The variable central
element in this case is the head of a monster which is associated with the month Cumku. The Initial Series, 13.0.0.0.0,
is followed by the Calendar Round date 4 Ahau 8 Cumku. It is
of interest that the coefficient of the uinal period is the
normal zero while a variant form of the zero glyph appears
with the katun, tun, and kin periods.

A more typical example of an Initial Series is shown in
figure 11.11. This inscription, from Yaxchilan, may be transcribed as follows.

A1	Initial Series introductory glyph [the central element is a monster head associated with the month Zip]
B1-B3	9.0.19.2.4
A4	2 Kan
B4	"the 8th Lord of the Night is in office"
A5	[significance unknown]
B5	"the moon is 7 days (or nights) old"
A6	"the moon is the 3rd in the lunar semester"
B6a	[significance unknown]
B6b	"the lunation has 29 days"
A7a	2 Yax

Fig. 11.11. The Initial Series and Supplementary Series on Lintel 21 at Yaxchilan (redrawn from Graham and Von Euw 1977-1979).

The above Initial Series runs from B1 to B3 and is counted from the unstated base at 4 Ahau 8 Cumku to reach the terminal date 2 Kan 2 Yax recorded at A4 and A7a. Between the separated parts of the Calendar Round date is a sequence of glyphs, often found with Initial Series, known as a "Supplementary Series". The glyph at B4 marks a station in a 9-day cycle referred to as the "Lords of the Night" cycle. Since the tun of 360 days is divisible by 9 the proper station in the 9-day cycle can be calculated from the uinal and kin counts by the formula $2n_2 + n_1$, mod 9. The glyph at B5 states the age of the current moon while that at A6 signifies the position of the moon in a lunar semester of 6 moons. At B6b is a statement that the current (or possibly the immediately preceding) lunation was of 29 days duration.

POSITIONAL NOTATION

In the codices the usual method of recording chronological counts was by means of a system of positional notation not requiring period glyphs. This notation consisted of a vertical arrangement of numbers in which the lowest position was used to represent the number of kins and successively higher positions were used to represent the number of each of the successively larger periods constituting the count. As a result, the Maya used position to distinguish among the tun, uinal, and kin counts as well as to distinguish among the place-values of the tun count.

Many examples of chronological counts expressed in positional notation are found on page 24 of the Dresden Codex illustrated in figure 11.12. A transcription of the page is shown in Table 11.3. The page in question serves to introduce a five page Venus table which associates four selected points of the Venus cycle to specified Calendar Round

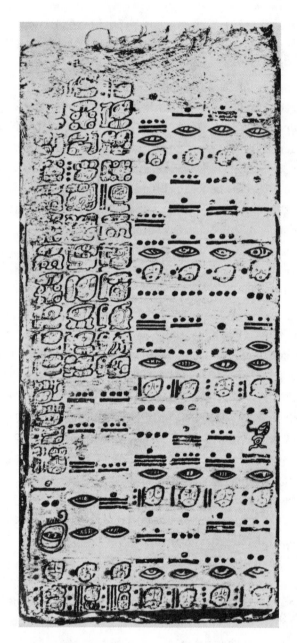

Fig. 11.12. Page 24 of the Dresden Codex (from Thompson 1972, courtesy of the American Philosophical Society).

Table 11.3 — Columns A, B, C (left portion):

#	A	B	C
1	1 Ahau	18 Kayab	1 Ahau
2			8 Cumku
3		Venus	
4	East		God L
5	Venus	Regent$_{48}$	Venus
6	Venus	Regent$_{49}$	Lahun Chan
7	Venus	Regent$_{50}$	Venus
8	Venus	Regent$_{46}$	Victim$_{46}$
9	Venus	Regent$_{47}$	Victim$_{47}$
10			Victim$_{48}$
11			Victim$_{49}$
12			Victim$_{50}$
13			
14			
15			
16		9	9
	RN	9	9
	6	16	9
	2	0	16
	0	0	0
	4 Ahau	1 Ahau	1 Ahau
	8 Cumku	18 Kayab	18 Uo

Right portion (four numeric columns):

1			
1	15	10	5
1	16	10	5
14	6	16	8
0	0	0	0
1 Ahau	1 Ahau	1 Ahau	1 Ahau
1			
5	9	4	1
14	11	12	5
4	7	8	5
0	0	0	0
1 Ahau	1 Ahau	1 Ahau	1 Ahau
4	4	4	3
17	9	1	13
6	4	2	0
0	0	0	0
6 Ahau	11 Ahau	3 Ahau	8 Ahau
3	2	2	2
4	16	8	0
16	14	12	10
0	0	0	0
13 Ahau	5 Ahau	10 Ahau	2 Ahau
1	1		
12	4	16	8
5**	6	4	2
0	0	0	0
7 Ahau	12 Ahau	4 Ahau	8* Ahau

* Should be 9

** Should be 8

Table 11.3. A partial transcription of page 24 of the Dresden Codex.

positions. The Venus table contains thirteen rows of Sacred
Round dates, each row running across five pages and accommo-
dating five Venus cycles, one cycle per page. These Sacred
Round dates can be tied into the Calendar Round in three ways
by using any one of three available rows of Vague Year dates.
The whole-day mean for synodic Venus revolutions is 584 days
and this is the period underlying the construction of the
Venus table. Because of its structure, each row of the table
spans a chronological interval of $5 \times 584 = 2920$ days or in
Maya notation a count of 8.2.0. The cumulative interval
spanned by succeeding rows of the table from the first on is
equal to successive multiples of 8.2.0 up to the thirteenth.
Since $5 \times 584 = 8 \times 365$, the $13 \times 5 = 65$ Venus cycles in the
table comprise $13 \times 8 = 104$ Vague Years or 2 Calendar Rounds.

Returning to Dresden 24, it can be seen that multiples of
8.2.0 are recorded in positional notation beginning in the
lower right corner, proceeding from right to left through
four columns and continuing in the same manner through the
second and third tiers from the bottom. In all, there are 12
multiples recorded in this part of Dresden 24 rising from
8.2.0. to 4.17.6.0. The 13th multiple, 5.5.8.0, equal to 2
Calendar Round intervals is written in the upper right corner
but is partially obliterated. The other entries in the upper
tier of numbers can be reconstructed to yield 4, 6, and 8
Calendar Round intervals as indicated in Table 11.3.

As is evident in figure 11.12, a zero variant is employed
which differs from that used in the inscriptions. It is a
stylized shell always painted red. A more elaborate version
of the shell glyph for zero is used to record the number of
tuns in the 5th multiple of 8.2.0, that is 2.0.10.0.

Following each of the multiples of 8.2.0 on Dresden 24 is
the Sacred Round date which is reached when the given

multiple is counted from 1 Ahau. The latter date occurs frequently on this page and functions as a ritual base for Venus
calculations. The smallest multiple is assigned an erroneous
date. Indeed, if one counts from 1 Ahau through an interval
of 8.2.0 one reaches 9 Ahau rather than 8 Ahau as the Maya
scribe has written. The error is an interesting one because
8.2.0 ≡ 8, mod 13, and consequently, the number 8 can be used
to generate successive trecena coefficients in the table of
multiples. Thus, the scribal error may reflect a procedure
for constructing the table in which the number 8 played a
special role. A second error appears in the record of the
4th multiple where 1.12.5.0 is shown rather than the required
1.12.8.0.

The fourth tier of numbers, not yet discussed, are all
multiples of 260 days and hence preserve the date 1 Ahau as
do the numbers in the upper tier. However, unlike the numbers in the upper tier, these are not multiples of the Venus
cycle. They are calculation factors which can be used to fix
the astronomical alignment of the Venus table with respect to
its chronological base at 9.9.9.16.0, 1 Ahau 18 Kayab.[8] This
chronological base is specified by an Initial Series in the
third column on Dresden 24. Its corresponding Calendar Round
date appears at the bottom of the second column.

Each page in the Venus table contains three illustrations.
The upper picture depicts what are believed to be presiding
deities associated with the various Venus cycles and referred
to as "Venus regents". The names of these regents are found
in the second column of the glyphic commentary on Dresden 24
and are noted in Table 11.3. Each of the name glyphs is preceded by a glyph signifying the planet Venus. The middle
picture on each of the Venus pages depicts a spearthrower

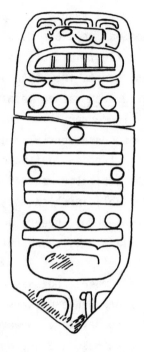

Fig. 11.13. Stela 1 at Pestac (after Blom 1935).

representing a manifestation of Venus at heliacal rising and the bottom picture represents victims of these spearthrowers. The names of two of the manifestations of Venus, God L and Lahun Chan, followed by Venus titles, are found in the third column of Dresden 24 and are noted in Table 11.3 as are the five victims of heliacal risings of Venus named in the closing glyphs of the same column.

The use of positional notation is rare in the monumental inscriptions but is not entirely absent. A Maya stela from Pestac, shown in figure 11.13, has an example of an Initial Series expressed in this system. It reads 9.11.12.9.0 and leads to a poorly preserved date 1 Ahau 8 Cumku.

Despite the scarcity of Maya inscriptions employing positional notation, the system does have impressive antiquity. The earliest known monument which may be of Maya origin and which exhibits this type of notation is Stela 5 from Abaj Takalik. The monument, found in 1976, bears a probable Initial Series record of 8.4.5.17.11 corresponding to A.D. 126 (Graham 1977).

However, there are even older monuments combining both the Long Count chronology and positional notation which predate those of undisputed Maya origin. One of the oldest and best preserved, Stela C from Tres Zapotes, is illustrated in figure 11.14 (Coe 1976). The monument displays an Initial Series of 7.16.6.16.18 corresponding to 31 B.C. This chronological record has an interesting modern history. The bottom part was unearthed in 1939 at which time the series of numbers on it was interpreted as the lower portion of an Initial Series. This demanded a missing baktun coefficient of 7. That reading was challenged because it presumed a Maya style Initial Series while the monument appeared to be of non-Maya origin. Also, the date was believed to be much too early for

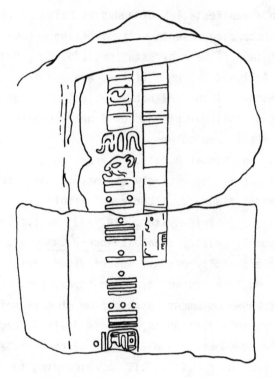

Fig. 11.14. Stela C at Tres Zapotes (after Coe 1976).

known Mesoamerican cultures. Nevertheless, as the archaeological horizon in Mesoamerica was pushed further and further
back the original reading found its defenders. Any doubt as
to its correctness was finally dispelled in 1973 when the
upper part of the stela was discovered in the same location
that had yielded the lower part 34 years earlier. Not only
did the new found fragment show that the baktun coefficient
was 7, it also had an early form of the Initial Series introductory glyph.

AN ALMANAC FROM THE DRESDEN CODEX

The Dresden Codex features many divinatory almanacs possessing a calendrical structure which is based on a partition
of the Sacred Round. This is achieved by specifying a set of
dates which divide the 260-day cycle into subsections. Each
subsection usually contains a brief text and an associated
picture. Usually, these almanacs are based on regular parti-
tions, the most common being a division into fifths with the
dates separated by intervals of 52 days. Other common divisions are into fourths with subintervals of 65 days and into
tenths with subintervals of 26 days each. Of mathematical
interest is the abbreviated quasi-tabular format in which the
calendrical structure of these almanacs is presented. Day
names are suppressed to economize space but the trecena coefficients are provided by red numerals while the intervening
Distance Numbers are expressed by black numerals.

A fairly typical divinatory almanac, from Dresden 13b-14b,
is reproduced in figure 11.15. The calendrical portion of
the almanac opens with a column of day signs preceded by a
red trecena coefficient of 6. The day glyphs can be identified from top to bottom as Ahau, Eb, Kan, Cib, and Lamat. A
row of numbers in which the colors alternate from black to

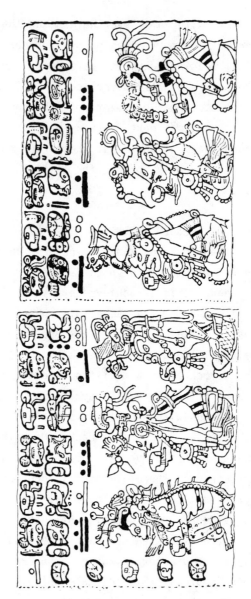

Fig. 11.15. Pages 13b–14b of the Dresden Codex (drawing from Villacorta and Villacorta 1930).

	DN	Section 1	DN	Section 2	DN	Section 3	DN	Section 4	DN	Section 5	DN	Section 6
6	13	6	9	2	7	9	7	3	7	10	9	6
Ahau		Ben		Ik		Muluc		Cib		Akbal		Eb
Eb		Chicchan		Ix		Imix		Lamat		Men		Kan
Kan		Caban		Cimi		Ben		Ahau		Manik		Cib
Cib		Muluc		Etz'nab		Chicchan		Eb		Cauac		Lamat
Lamat		Imix		Oc		Caban		Kan		Chuen		Ahau

Table 11.4. The calendrical structure of the almanac on pages 13b-14b of the Dresden Codex.

red is also present. The almanac opens with the date 6 Ahau
in the left column. The first black number, representing a
Distance Number of 13 days, leads to the date 6 Ben of which
only the red trecena coefficient is recorded. The first text
of four glyphs above these numbers and the picture below are
associated with that date. One then proceeds to the next
section of the almanac by counting an additional 9 days,
given by the second black number, to reach 2 Ik. Once again,
only the red trecena coefficient is shown. One proceeds in
this manner through the third, fourth, and fifth sections of
the almanac to arrive at the last section associated with the
date 6 Eb. Now, the red 6 at the top of the opening column
is intended to apply to all the day glyphs below it. Thus,
the date 6 Eb is found there, immediately beneath 6 Ahau.
This signals that one is to proceed through the almanac a
second time beginning with 6 Eb. After five passages through
the almanac, one ultimately reaches the terminal date 6 Ahau.
In agreement with the cyclical nature of divinatory almanacs,
this coincides with the beginning date. A complete description of the calendrical data in the almanac is given in Table
11.4 by means of a more conventional tabular arrangement.

The almanac under study is concerned with maize, an agricultural staple for the Maya. The glyphic commentary associated with each section follows a formulaic pattern -- action,
object, protagonist, augury -- common to many of the short
texts in the Dresden Codex. THe first glyph, in each case,
is a verb which signifies the action taking place in the corresponding section. The main sign of the verb is found in
all six sections, but in the first three it takes a common
prefix and lacks a postfix while in the last three it lacks a
prefix and takes a common postfix. The verb has been read as
a phonetic compound ma-ch(a), for mach, 'to take in one's

hand, to hold' (Fox and Justeson 1984, pp.21-22). The second
glyph in each section has as its main sign a kan glyph (see
the glyph for the day Kan) signifying maize. As happened
with the first glyph, the affixes of the second vary according as to whether it is found in the first three sections or
the last three. This phenomenon indicates that, although the
two halves of the almanac deal with substantially the same
thing, some modification takes place in the action and the
object described in the two parts of the almanac. It has
been suggested that the first three sections refer to an
action which influences the young shoots of the new maize
while the last three refer to a similar action which influences the matured maize (Thompson 1972, pp.40-41). The
maize, symbolized by a kan glyph, can also be seen resting in
the hand or on the fingers of the deity pictured in each section. The name glyph of each of these deities is given in
the third position of the corresponding text. In order of
appearance, these are the death god, the maize god, God C (a
popular deity whose identity is unknown), God L (a Venus god
and god of the underworld), God Q (a god of sacrifice), and
Itzamna (a creator god). The final glyph in each section
gives the augury for the date as it relates to the given
clause. The auguries, from left to right, may be loosely
translated (with quotes) or classified (without quotes) as
"death", favorable, "very good", favorable, "evil", and
favorable. It is clear that these auguries result from the
character of the gods acting as protagonists in each section.

Drawing on the above comments, it is possible to capture
the essential meaning of the glyphic texts in the almanac,
although a precise translation of some of the content is
still in question. For example, the first section depicting

the death god holding the maize sign has the glyphic caption
"[He] takes in his hand -- the maize -- the death god --
death" while the second section depicting the maize god holding the maize sign has the caption "[He] takes in his hand --
the maize -- the maize god -- good fortune".

HEAD VARIANT NUMERALS

Of all the numerals devised by man none can compare for
sheer beauty with the Maya head variant numerals. These are
represented by portraits of heads whose features or attributes are the key to the number thus portrayed. Most of the
numerical profiles have been identified as those of gods and
it is safe to assume that all have a similar derivation. The
head variant glyphs of the numbers from 1 to 19, as well as
that for 0, are represented in figure 11.16. Although the
head variants were depicted with considerable artistic free-
dom, there are characteristic elements which are usually
found in all the varied forms for a given number. A list of
the identities of the numeral portraits, where known, and of
their defining characteristics is given below.[9]

1 **Young earth goddess.** The head is that of a youthful
female distinguished by a lock of hair which passes in front
of the ear, and curves forward along the base of the jaw. It
usually has a small ornament on the forehead and an IL sign
frequently appears on the cheek.

2 **God of sacrifice?** The numerical profile is characterized by a hand which surmounts the head.

3 **God of wind and rain.** The head is that of a youthful
god distinguished by an ik symbol (see the glyph for the day
Ik) usually worn on an ear flap, a disk edged with circlets
which is set on the forehead, and a banded headdress.

Fig. 11.16. Head variant numerals.

Sometimes an IL mark appears on the cheek.

4 <u>Sun god</u>. The head of the aged sun god can be recognized by a large squarish eye with pupil set in an inner corner, a roman nose, upper incisors filed to a T-shape, and frequently an incised <u>kin,</u> 'sun', glyph set into the face.

5 <u>Aged god of the underworld</u>. The head of this old god may be identified by a <u>tun</u> sign headdress.

6 <u>Rain god?</u> The portrait is characterized by a hafted axe set in a large squarish eye.

7 <u>Jaguar god of the underworld</u>. The deity head is distinguished by a loop which passes under the eyes and which is loosely tied over the bridge of the nose. The eyes are squarish, the nose is Roman and the central incisors of the upper jaw are filed to a T-shape.

8 <u>Maize god</u>. The head is youthful and has a maize plant or more frequently a spiral on the forehead. Usually the head has maize foliage on the side of the face falling in front of, over, or behind the ear and reaching to the chin.

9 <u>Serpent god?</u> The god is easily identified by the presence of dots, probably jaguar spots, near the mouth. Often the face is bearded and a <u>yax</u> symbol (see the first element in the glyph for the month Yaxkin) is on the forehead.

10 <u>Death god</u>. The principal characteristic of the death god's head is a fleshless jawbone. Other diagnostic traits include a fleshless nose, a "percentage" sign on the cheek, three dots on the upper part of the head, and a "death eye" on the forehead.

11 <u>Earth god</u>. The numeral head is marked with a <u>caban</u> sign (see the glyph for the day <u>Caban</u>) which signifies "earth".

12 <u>Identity unknown</u>. The head for 12 does not have a recognizable invariant feature. Sometimes it is depicted with a

sky sign (see illustration) headddress.

13 Reptilian monster. The most characteristic attribute
of this head is a long pendulous nose.

Even though the Maya had a distinct head variant which
could represent the number 13, they more often used an alternative form consisting of a blending of the heads for 3 and
10. The head for 3 was combined with the fleshless jawbone
of 10 to yield a composite head having the value 3 + 10 = 13.
The head variants for the numbers 14 to 19 were expressed in
a similar way by adding the bared jawbone of the head for 10
to the corresponding head variants for the numbers 4 to 9.
These formations parallel the construction of the Maya number
words from 13 to 19 by addition to a base of 10.

The portrait glyph for 0 is distinguished by the presence
of a hand across the lower jaw.

The use of head variant numerals is not restricted to any
one context nor to any one Maya site. They are found in Cal-
lendar Round dates and Distance Numbers but most often in
Initial Series. Examples of dates, all from Palenque, which
employ head variant numerals are shown in figure 11.17, a-c.
The first date is 8 Ahau 18 Zec and allows one to examine the
similarities and differences between the related head variants for 8 and 18. The second example shows the date 1 Ahau
13 Mac in which only the coefficient of the Sacred Round portion is honored with a head variant numeral. The third example has two head variants for 5 and records the date 5 Eb 5
Kayab. Another example, also from Palenque, in figure
11.17d, illustrates a Distance Number of 2.2.14.5. The Distance Number combines head variant numerals for the tun count
with bar and dot numerals for the uinal and kin counts.

An Initial Series from Quirigua which shows an extensive

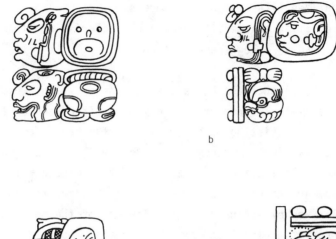

a   b

c   d

Fig. 11.17. Calendar Round dates and a Distance Number employing head variant numerals: (a) 8 Ahau 18 Zec, Palenque, Tablet of the Cross, A8-B9; (b) 1 Ahau 13 Mac, Palenque, Tablet of the Foliated Cross, A8-A9; (c) 5 Eb 5 Kayab, Palenque, Temple of the Cross, Door jamb panel, B3x-A4x; (d) 2.2.14.5, Palenque, Tablet of the 96 Glyphs, F7-F8.

use of head variant numerals is depicted in figure 11.18.
The text may be transcribed as follows.

A1-B2	Initial Series introductory glyph [the central element is a monster head associated with the month Zip]
A3-A5	9.16.10.0.0
B5	1 Ahau
A6	"the 9th Lord of the Night is in office"
B6	"the moon is being born"
A7	"the moon is the 6th in the lunar semester"
B7	[significance unknown]
A8	"the lunation has 30 days"
B8	3 Zip

The Initial Series of 9.16.10.0.0, counted from the chronological base date at 4 Ahau 8 Cumku, leads to the terminal
date 1 Ahau 3 Zip. As is often the case, a Supplementary
Series intervenes between the two parts of the Calendar Round
date.

In several inscriptions, the Maya used numerals which were
even more elaborate than the head variants. These consisted
of complete anthropomorphic figures, usually characterized by
portraying their heads in the same manner as the corresponding head variant numerals. In one exceptional case, in the
Initial Series from the Palace Tablet at Palenque, numbers
were recorded by full figure glyphs having identifying marks
on the limbs of the body or having characteristic symbols set
into their headdresses. These particular numerals, illustrated in context in figure 11.19, rank among the most beautiful ever executed and are depicted with superlative grace
and dignity.

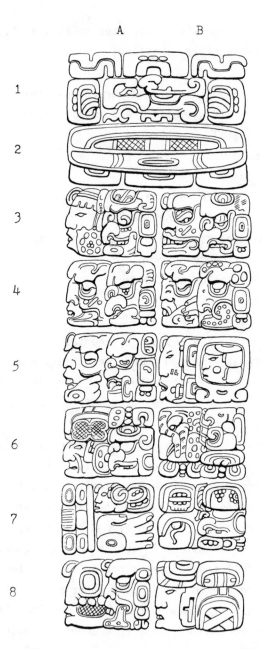

Fig. 11.18. The east side of the Stela F at Quirigua, A1-B8 (from Maudslay 1889-1902).

Fig. 11.19. The Initial Series on the Palace Tablet at Palenque.

The Initial Series opens with an introductory glyph in
which the central figure has ik markings on the limbs identifying him as patron of the month Mac and anticipates the
month in which the terminal date of the Initial Series falls.
The baktun glyph is a full figure avian variant with a characteristic hand across the lower jaw. Its numerical coefficient has jaguar markings on the limbs, a jaguar ear, set
above the human ear and a yax sign set into the headdress.
These elements identify the figure as the number 9. The
katun glyph is also represented by a full figure avian variant. Its coefficient has no markings on the limbs but wears
a death head pectoral and a skull headdress with fleshless
jawbone, indicating that it signifies the number 10. The tun
glyph is another full figure avian variant and is typified by
a fleshless jawbone. Its coefficient has caban markings on
the limbs identifying it as the number 11. The uinal glyph
is rendered by a full figure frog. The corresponding numerical coefficient has no signs on the limbs but wears a skull
headdress to which is attached a cascade of feathers and a
jaguar's ear. These attributes, together with its occurrence
as the coefficient of the uinal period (which excludes the
possible value of 19), suggest that it has the value 17. The
kin variant is an anthropomorphic figure with an extra deer-
like ear set into the hair above the normal ear. Its coefficient has zero symbols on the arm and represents the number
0. Finally, the Sacred Round date has a spider monkey within
the day sign cartouche and a coefficient whose limbs are
marked with the caban signs characteristic of the number 11.
Since the kin coefficient is zero the date can only be 11
Ahau. The sundry data are sufficient to determine the Initial Series as 9.10.11.17.0 and its terminal date as 11 Ahau
(8 Mac).

A HIEROGLYPH FOR 20

In many of the ritual almanacs of the Dresden Codex, Distance Numbers of not less than 20 and not more than 39 days are expressed by a special day count notation. In this system a moon sign is used to represent the number 20. Numbers above 20 are represented by appending a bar and dot numeral of sufficient magnitude to the moon sign. The composition of the moon sign and the bar and dot numeral is always additive. It yields a count of days alone rather than separate counts of uinals and kins. An example of a count of 20 + 11 = 31 days is shown in figure 11.20a. The glyphic expression may be translated as hun kal (catac) buluc, "one score (and) eleven". Sometimes the bar and dot portion may be written horizontally as in the case of the 20 + 6 = 26 days in figure 11.20b. A singular example illustrating a vertical placement of the moon sign and bar and dot numeral in a Distance Number of 6 + 20 = 26 days is shown in figure 11.20c. A somewhat different and infrequent manner of expressing this type of day count is found in figure 11.20d which depicts a count of 16 + 20 = 36 days. In this case a glyph having the phonetic value tu appears above the moon sign. The expression may be translated as uaclahun tu kal, "sixteen in the (second) score", and reflects a more colloquial method of numeral formation than that used in the first two examples.

A different form of the moon sign appears in the inscriptions as a symbol for 20. It is sometimes used in Distance Numbers as in the count of 11 + 20 = 31 days shown in figure 11.20e. An interesting combination of the normal zero sign with the glyph for 20 can be seen in the Distance Number of 0 + 20 = 20 days illustrated in figure 11.20f. The only example of the use of the moon sign for 20 in a count exceeding 39 days is in the Distance Number of 2 tuns and 36 days

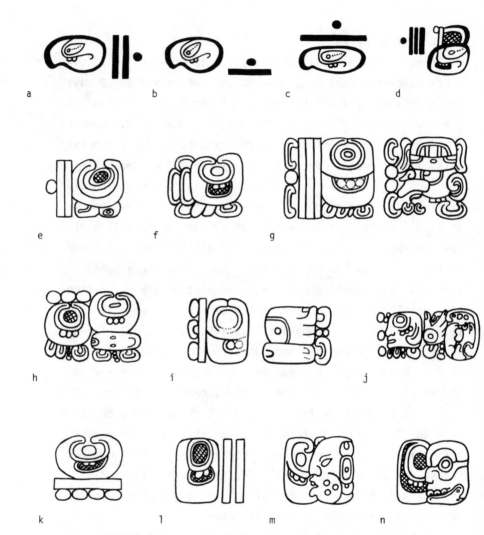

Fig. 11.20. Chronological counts which employ the moon glyph for 20.
Distance Numbers: (a) 31 days, Dresden Codex, 15b; (b) 26 days, Dresden
Codex, 13a; (c) 26 days, Dresden Codex, 8b; (d) 36 days, Dresden Codex, 27c;
(e) 31 days, Balakbal, Stela 5, D10; (f) 20 days, Palenque, Tablet of the Cross,
B13; (g) 2 tuns and 36 days, Tikal, Stela 22, B9-A10.
Moon age counts: (h) 23 days, Piedras Negras, Stela 12, A9; (i) 27 days,
Naranjo, Stela 13, F5-E6; (j) 23 days, Quirigua, Monument G, I2-J2.
Length of lunation counts: (k) 29 days, Palenque, Tablet of the Cross, A13; (l) 30
days, Piedras Negras, Stela 1, B2; (m) 29 days, Quirigua, Stela J, A13; (n) 30
days, Piedras Negras, Lintel 3, F1.

shown in figure 11.20g.

The moon glyph for 20 is also employed in a different type of chronological format. In those cases where a Maya text begins with an Initial Series it is common to find supplementary information pertaining to its terminal date. This typically includes a reference to the current age of the moon. If the moon age at the date in question exceeded 19 days it was recorded by a count of days (or possibly, nights) employing the moon glyph for 20. Examples of moon-age counts, of 23 and 27 days, respectively, are shown in figure 11.20, h-i. A moon-age count of 23 days in which the moon glyph for 20 is a head variant is illustrated in figure 11.20j. The glyphic context in which moon-age counts occur may be seen in figure 11.11, B5, where a moon-age count of 7 days is recorded.

The moon glyph for 20 also appears in a glyphic expression indicating whether the current (or possibly, previous) lunation was of 29 or 30 days duration. This compound consists of the moon glyph with a coefficient of 9 or 10 attached as a postfix, unlike the previous situation where coefficients were prefixed. Examples of length of lunation glyphs representing 29-day and 30-day moons, respectively, are shown in figure 11.20, k-1. In some instances, the moon glyph for 20 is depicted in a half-moon form as may be seen in figure 11.20m where a lunation of 29 days, employing a head variant for 9, is represented. Another example, this time of a 30-day lunation and employing a head variant for 10, is shown in figure 11.20n. The usual context in which the length of a lunation is found may be observed in figure 11.11, B6b, where a 29-day moon is recorded and again in figure 11.18, A8, where a 30-day moon is recorded by means of head variant numerals for 20 and 10.

A singular usage of the moon glyph for 20 in an Initial

Series occurs on Stela 5 at Pixoy (Closs 1978). It appears
as the coefficient of the katun term. The Initial Series and
its terminal date read 9.13.20.0.0, 6 Ahau 13 Muan, a unique
but correct rendering of the expected 9.14.0.0.0, 6 Ahau 13
Muan.

SUPPLEMENTARY NUMERAL VARIANTS

In addition to the many numeral variants already considered, the Maya sometimes used still other glyphic forms to
record numbers. These additional variants are discussed here
separately, because of their rare, sometimes singular, occurrences. Despite their scarcity of use, they can be recognized as legitimate numerals because of the context in which
they are found. For example, the three glyphs illustrated in
figure 11.21, a-c, are constituent terms of different Initial
Series. In each case, it is evident from the context that
the glyph consists of the indicated period and a numerical
coefficient having value 0. Thus, the prefixes of these
glyphs are known to be distinctive variants of zero.[10]

The number 1 is at times represented by a finger. A nice
example of this usage can be seen in figure 11.21d which
exhibits a Distance Number of 1.1.17. The numerical coefficient to the tun is a finger having value 1.

A unique head variant for 3, depicted in figure 11.21e,
appears in the Dresden Codex. It is composed of a cursive
Ahau surrounded by U-shaped elements and a superfix which
includes a disk formed by a circle of dots. The last feature
is frequently found in the headdress of the head variant for
3 used in the inscriptions. The present example is found at
the head of a column of day signs and occupies a position
which demands the numeral 3.

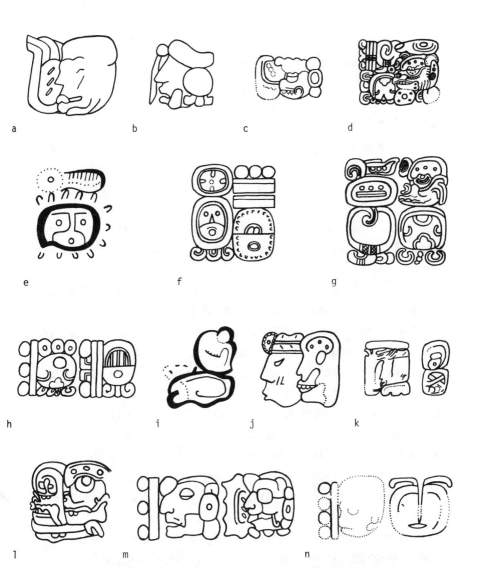

Fig. 11.21. Supplementary numeral variants: (a) 0 <u>kins</u>, Sacchana, Stela 2, A5; (b) 0 <u>tuns</u>, Altar de Sacrificios, Stela 9, CD₂; (c) 0 <u>uinals</u>, Pixoy, Stela 5, B4; (d) 1.1.17, Naranjo, Hieroglyphic Stairway, M3; (e) 3, Dresden Codex, 9b; (f) 4 Ahau 13 Mol, Copan, Stela 1, D6; (g) 10.8, Copan, Stela I, C6; (h) 12.9.8, Palenque, Temple of the Inscriptions, West Panel, E1-F1; (i) <u>bu-lu-c(u)</u> = 11, Dresden Codex, 19a; (j) 13, Tila, Stela B, A4; (k) 15 Uo, Chichén Itzá, Las Monjas, Lintel 6, A3-A4a; (l) 16, Xcalumkin, all-glyphic column, A4; (m) 18 Yax, Tikal, Stela 6, A10; (n) 19 Pax, Yaxchilan, Lintel 47, A3-B3.

It will be recalled that the head variant for the number 4 is the head of the sun god which is usually incised with a kin glyph. On one occasion the kin glyph itself substitutes for the number 4. This occurs in the Calendar Round date 4 Ahau 13 Mol shown in figure 11.21f.

An elaboration of the bar and dot numeral for 8 is found in the Distance Number of 10.8 shown in figure 11.21g. The uinal coefficient is a head variant 10 and the kin coefficient is a bar and dot 8 inside an oval surmounted by maize foliage. The peculiar treatment of this numeral is in agreement with the fact that the deity of the number 8 is the maize god.

A unique representation of the number 9 as three 3's may appear in the Distance Number of 12.9.8 shown in figure 11.21h. The uinal coefficient consists of three circles and it has been suggested that they might be equivalent to the disk characteristic of the head variant for 3 (Teeple 1931, pp.81-82). In fact, the tiny circles expected around the outer edge of the disks are not visible, possibly due to erosion but also possibly due to the fact that they never existed. After examining the artifact containing this Distance Number, Linda Schele (personal communication November, 1985) has opted for the latter position. Thus, the exact significance of this record is questionable.

There is an unusual usage of a phonetic construction for the number 11, buluc, in the Dresden Codex (Kelley 1976, p.171). The written compound, illustrated in figure 11.21i, appears at the top of a column of day signs where a numeral 11 is demanded. Of the three components of the glyph, the first has been destroyed, but the remaining two are known to have the phonetic values lu and cu. It appears that the compound was to be read as bu-lu-c(u), '11', where the final vowel sound was to be dropped, a common procedure in Maya syllabic writing.

It has been observed that there were two types of head variants for the number 13. One consisted of a distinctive deity head with a long pendulous nose and the other was formed by joining the head for 3 with the fleshless jawbone of the head for 10. There is also a singular example, shown in figure 11.21j, of the number 13 recorded by tandem heads for 3 and 10.

A combination of a numerical bar for 5 and a head for 10 to yield a coefficient of 15 occurs in the Vague Year date illustrated in figure 11.21k. Although the head is poorly preserved and its prefix is indistinct, a numerical bar for 5 is evident. From related texts it is known that this date must record 15 Uo.

There is one instance, shown in figure 11.21l, of the number 16 being represented by tandem heads for 6 and 10 (Closs 1983). The formation is similar to that in the singular construction of 13 discussed above.

Two other examples exhibiting a mixture of bar and dot numerals and the head variant for 10 are shown in figure 11.21, m-n. They are both found in Vague Year dates, the first being 18 Yax and the second 19 Pax.

RING NUMBERS

Several Distance Numbers in the Dresden Codex have their kin coefficients enclosed in a conspicuous red loop tied with a knot at the top. This feature has given rise to the term "Ring Number" to describe a Distance Number of this type. The Ring Numbers to be discussed here are always accompanied by another Distance Number called a "Companion Number" and are always followed by the base date of Maya chronology, 4

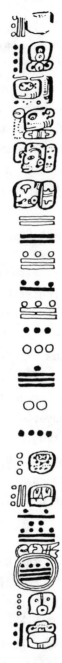

Fig. 11.22. Page 63, Column C of the Dresden Codex (drawing from Villacorta and Villacorta 1930).

Ahau 8 Cumku. There are rare usages of Ring Numbers in other
contexts but they shall not be considered here (Thompson
1972, pp.21-22).

An example of a Ring Number of 7.2.14.19 is illustrated in
figure 11.22 near the bottom of the column immediately above
the base date 4 Ahau 8 Cumku. Above it appear two Sacred
Round dates and above these appear two Companion Numbers, one
red and one black. The colors permitted the Maya to record
two Distance Numbers, without confusion, in a space which
would otherwise only accommodate one. At the top of the column is the Calendar Round date 13 Imix 9 Uo followed by a
brief four glyph text. It is unusual to find more than one
Companion Number associated with the same Ring Number but
otherwise the chronological structure of this passage is typical. If one employs the red companion number, the content
of the column can be roughly expressed as follows.

On 13 Imix 9 Uo a specified event took place in which the
rain god Chac was the protagonist. (The nature of this event
has not been determined but it is also referred to in an 819-
day augural count appearing in dynastic texts.) From this
date, a Distance Number of 10.13.13.3.2 leads to 13 Akbal (1
Kankin), the Vague Year portion of the date not being listed
because of a lack of space. The Distance Number in question
was counted from a chronological position 7.2.14.19 prior to
the zero date of the Long Count at 4 Ahau 8 Cumku.

It is evident from the above relationships that the Long
Count position of the initial date 13 Imix 9 Uo is fixed by
the Ring Number. In this respect, the Ring Number may be
regarded as a negative Initial Series that records the proper
position of the initial date in the Long Count. Therefore,
if a Distance Number is counted from that date, the Long
Count position of the date reached can be obtained from the

algebraic sum of the Distance Number and the Initial Series belonging to the initial date. In the present case, the calculation, namely

$$10.13.13.3.2 - 7.2.14.19 = 10.6.10.6.3$$

yields the Long Count position (or Initial Series record) of the date 13 Akbal 1 Kankin.

From another point of view, the Ring Number may be considered to be an ordinary Distance Number which leads forward in time from the initial date to the zero date 4 Ahau 8 Cumku. This raises the question of whether a Ring Number is best interpreted as a negative Initial Series or as a positive Distance Number. If the Maya perceived Ring Numbers as ordinary Distance Numbers one would expect them to be represented in the same fashion as other Distance Numbers. This is not the case. The extraordinary notation devised by the Maya to write Ring Numbers suggests that they were perceived as something more. This supports the notion that a Ring Number was intended to provide a statement of the Long Count position of an initial date, that is, it constituted a negative Initial Series.

The Ring Number passage in figure 11.22 may be summarized in the form shown below.

	Dresden 63C	Long Count Position
Initial Date	13 Imix 9 Uo	-7. 2.14.19
Companion Number 1 (black)	10.8*. 3.16. 4	
Companion Number 2 (red)	10.13.13. 3. 2	
Terminal Date 1	3 Chicchan (8 Zac)	10. 6. 1. 1. 5
Terminal Date 2	3 Akbal (1 Kankin)	10. 6.10. 6. 3

```
Ring Number              7. 2.14.19
Zero Date                4 Ahau 8 Cumku        13. 0. 0. 0. 0
                         * should be 13
```

A second example of a Ring Number passage in the Dresden Codex can be seen in figure 11.12, Columns A-B. A summary of the passage is presented below.

	Dresden 24, A-B	Long Count Position
Initial Date	1 Ahau 18 Kayab	-6. 2.0
Companion Number	9.9.16.0.0	
Terminal Date	1 Ahau 18 Kayab	9.9.9.16.0
Ring Number	6.2.0	
Zero Date	4 Ahau 8 Cumku	13.0.0. 0.0

In this instance, the initial date at A1-B1 is largely effaced. The Companion Number is located in Column B with the terminal date below it. The Ring Number and the zero date which follows it are located in Column A. Unusually, the Long Count position of the terminal date is recorded by an Initial Series in Column C preceded by the zero date at C1-C2.

It is interesting to note that just as positive Initial Series in the Dresden Codex are generally preceded by 4 <u>Ahau</u> 8 <u>Cumku</u> so Ring Numbers are followed by it. In addition, it may be pointed out that the terminal dates of Ring Number calculations always fall in historical times whereas the initial dates fall in the mythological era preceding the base date of Maya chronology.

A regular, although not constant, feature of the Companion Numbers associated with Ring Numbers is their decomposability into a relatively large number of relatively low prime factors. For example, the Companion Numbers described above have the following prime factorizations:

Dresden 63C (black)   10.13. 3.16.4   = $2^2 \cdot 3^3 \cdot 61 \cdot 233$

Dresden 63C (red)   10.13.13. 3.2   = $2 \cdot 13 \cdot 59167$

Dresden 24B   9. 9.16. 0.0   = $2^5 \cdot 3^2 \cdot 5 \cdot 13 \cdot 73$.

The middle example is the unique instance of a Ring Number having such a large prime factor. The other two examples are much more typical and lead to the inference that these numbers are contrived (Lounsbury 1976).

It is very common to find that Companion Numbers include the factors $2^2 \cdot 5 \cdot 13 = 260$ as in the last example above. As a consequence, such a Distance Number will preserve Sacred Round positions. In the case of Dresden 24, the introductory page to the Venus table, 1 Ahau is used as a ritual base for Venus calculations, a result in keeping with the fact that 1 Ahau is also the calendar name for the Venus god. Thus, the Ring Number passage on Dresden 24 preserves this ritual base. In addition, the Companion Number on Dresden 24 includes the factors $2^3 \cdot 73 = 584$ and hence also preserves Venus positions in the canonical cycle for synodic revolutions of this planet. (The problem of preserving actual astronomical alignments was taken care of by calculation factors found in the fourth tier of numbers on the right half of Dresden 24.)

The Ring Number notation is not employed in the inscriptions but an analogue does occur on the Tablet of the Cross at Palenque. This text opens with a peculiar Initial Series which reads 12.19.13.4.0 and leads to the Calendar Round date 8 Ahau 18 Zec. The chronological position of this date is 6.14.0 prior to the zero date at 4 Ahau 8 Cumku. It has been commonly supposed that the Initial Series records a positive count of 12.19.13.4.0 measured from a base date 13 baktuns prior to 4 Ahau 8 Cumku. However, no substantial evidence confirming a Maya interest in this presumed chronological base has ever been presented. Moreover, every other known Initial Series is reckoned from 4 Ahau 8 Cumku and consistency demands that the same base be used in the present case. This can be achieved by interpreting the Long Count statement as a negative Initial Series equivalent to the Ring Number 6.14.0. The actual formula for the Initial Series can be derived from that of the zero date by the equation

$$13.0.0.0.0 - 6.14.0 = 12.19.13.40.$$

If the Initial Series from the Tablet of the Cross is an analogue of the Ring Numbers in the Dresden Codex, there should be an associated Distance Number possessing some of the numerical properties characteristic of Companion Numbers, leading from 8 Ahau 18 Zec to a later date in the historical era. No such Distance Number is recorded on the Tablet of the Cross but it is now known that such a Distance Number is implicit in the dynastic records of Palenque (Lounsbury 1976). Given that the Sacred Round position of the initial date is 8 Ahau and the expected preservation of Sacred Round positions by Companion Numbers, one looks for a similar date in the historic period at Palenque. Such a date exists and a very famous one too, namely, the birth date of Palenque's magnificent ruler Pacal. His birth date is 8 Ahau 13 Pop and its Long Count position, fixed by other texts at Palenque, is 9.8.9.13.0. The interval from the initial date on the Tablet of the Cross to Pacal's birth is then given by

$$9.8.9.13.0 + 6.14.0 = 9.8.16.9.0 .$$

This interval has the prime factorization

$$9.8.16.9.0 = 2^2 \cdot 3^2 \cdot 5 \cdot 7 \cdot 13 \cdot 83$$

and is clearly a contrived number possessing the characteristic numerical attributes one expects of Companion Numbers.

The relationship between the initial date on the Tablet of the Cross and the birth date of Pacal goes beyond the arithmetical. Indeed, the initial date marks the birth of an ancestral goddess who is the mother of the gods at Palenque. Since both dates are numerologically connected and mark similar events it can be inferred that the birth date of the goddess was selected so as to prefigure the birth of Pacal more than 3500 years later.

OTHER NOTATIONAL DEVICES

Many, but by no means all, Distance Numbers are preceded by a distinctive glyph which has been termed a "Distance Number Introductory Glyph". A typical such glyph is illustrated in figure 11.23a. The prefix is always selected from a small set of about a half-dozen members. The postfix, which is also a member of this group, is generally constant. The main sign, somewhat resembling a swastika, has the connotation of change. There are variants of the glyph similar to those in figure 11.23, b-c, where the main sign is replaced by a pair of glyphs which generally exhibit a natural opposition. Of the two examples which are shown, the first contrasts the caan, 'sky', and caban, 'earth', glyphs while the second contrasts the kin, 'day', and akab, 'night', glyphs.

The fact that the Distance Number Introductory Glyph signifies "change" and is followed by a Distance Number indicates that the change in question is chronological. Since the chronological change corresponds to the difference in Long Count positions between the initial and terminal dates of a calculation, the introductory glyph probably signifies a change or increment in the Long Count.

a   b   c

d   e   f   g   h

i

Fig. 11.23. Distance Number Introductory Glyphs: (a) Kuná-Lacanhá, Lintel 1, J5; (b) Copan, Temple 11, East Door, South Panel, B4; (c) Yaxchilan, Structure 33, Hieroglyphic Stairway, Block 7, H7.
Posterior date indicators: (d) Palenque, Temple of the Inscriptions, West Panel, S5; (e) Yaxchilan, Lintel 31, I3b; (f) Yaxchilan, Lintel 30, H4a.
Anterior date indicators: (g) Copan, Stela C, A7a; (h) Kuná-Lacanhá, Lintel 1, K2; (i) Palenque, Tablet of the 96 Glyphs, K2.

Frequently, special glyphs such as those shown in figure 11.23, d-h, are placed between a Distance Number and a following Calendar Round date. These serve to indicate whether the associated date is an initial or terminal date for the preceding Distance Number. For this reason they are called "anterior" or posterior" date indicators, respectively.

The prefix of the main sign in figure 11.23d only occurs when the associated date is the posterior date in the calculation, that is, when the preceding Distance Number counts from a previously precorded earlier date towards the associated date. Thus, the prefix characterizes the glyph as a posterior date indicator. The same prefix is also used with event glyphs in a similar capacity. Its presence in such a context indicates, that the associated event is the later of two events linked by a chronological count. An example of this usage may be seen in figure 11.8, R16-S16. David Stuart (personal communication, July, 1985) has amassed convincing evidence that the glyph in figure 11.23d may be translated as "then there came to pass", or more concisely "then occurred".

The example in figure 11.23e has a main sign consisting of the muluc glyph (see the glyph for the day Muluc) but carries the same affixes as the previous compound. Its characteristic prefix, identifies it as a posterior date indicator. Occasionally, that prefix is replaced by the u bracket seen in figure 11.23f. The latter compound can be seen in context in figure 11.7, H4a. Stuart has shown that the compounds with the muluc main sign are linguistically equivalent to the one in figure 11.23d and so may be read in the same way.

The glyphs in figure 11.23, g-h, employ the same main signs and postfix as the posterior date indicators but lack their determining prefixes and carry an extra postfix. Such compounds are found in situations where the associated date is the anterior date of the calculation, that is, when the preceding Distance Number is counted backwards from the associated date to a previously recorded later date. They can be translated as "after there came to pass", or "after occurred". An example of an anterior date indicator in context is found in the glyphic passage shown in figure 11.24. The sentence may be paraphrased as follows.

E10	[Its change was] 2 [kins], 12 uinals,
F10	10 tuns,
E11	6 katuns, and
F11	3 baktuns
E12	"after occurred"
F12	9 Ik [0 Yax];
E13	then was the birth of
F13	U-Kix-Chan
E14-F14	(titles of U-Kix-Chan)
E15	Lord of Palenque.

Note the prefix at E13 which indicates that the birth event is posterior to the preceding 9 Ik date. In this sentence, the terminal date of the calculation, that is the birth date, is not explicitly recorded. However, that date, 1 Kan 2 Cumku, can be recovered from the given data, here and elsewhere, in the inscription.

A different type of anterior date indicator, occurring at Palenque, consists of a wriggly creature similar to an eel or snake. When it is appended to a Sacred Round date, that date is the earlier of two dates linked by an associated Distance Number. An example of this type of date indicator is shown in figure 11.23i.

It was noted earlier that a "seating" glyph was sometimes

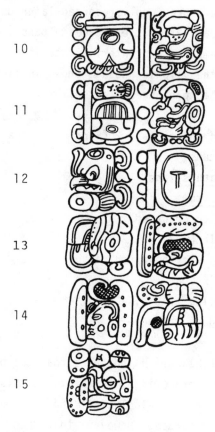

E   F

10

11

12

13

14

15

Fig. 11.24. A reference to the birth of the legendary U-Kix-Chan of Palenque; Palenque, Tablet of the Cross, E10-E15 (after Maudslay 1889-1902).

prefixed to a month glyph to express the installation day for
the given month (see figure 11.1, g-h). The same glyph also
appears as a prefix with another glyph which replaces the
normal tun glyph in certain circumstances. An example of a
"seating of the tun" expression is illustrated in figure
11.25a. This expression, most common at Palenque, refers to
the installation of a new katun in the Long Count, an event
which took place on the last day of the preceding katun. The
same expression is occasionally found on the 13 tun anniversary of an installation day, a tun ending which has the same
Sacred Round date as the installation day itself. The "seating of the tun" compound with the seating glyph merged with
the tun glyph and having a numerical coefficient of 4 appears
on the lid of Pacal's sarcophagus (see figure 11.9, Glyph 6).
As stated earlier, this expression refers to the four katuns
which were installed during the life of Pacal. An excellent
example of the "seating of the tun" glyph in context is
shown in figure 11.25b. The passage opens an inscription
from Palenque with a chronological statement fixing the
subsequent text in the Long Count. It may be paraphrased as
follows:

A1	[Its change was] 0 [kins], 7 uinals, and
B1	4 tuns
A2-B2	"from" 12 Ahau 8 Ceh
A3	[at] the seating of the tun;
B3	"then occurred"
A4-B4	6 Ahau 8 Cumku.

Note the use of the snake-like anterior date indicator at A2
and of a posterior date indicator at B3. The "seating of the
tun" at A3 describes the installation of a new katun (or its

A   B

1

2

3

4

a

Fig. 11.25. (a) Seating of the <u>tun</u>, Palenque, Temple of the Inscriptions, West Panel, C2. (b) Palenque, Dumbarton Oaks Relief Panel 2, A1-B4 (after Coe and Benson 1966).

13-tun anniversary) on 12 Ahau 8 Ceh and fixes the Long Count
position of that date as 9.11.0.0.0. The Long Count position, of the second date is then given by

$$9.11.0.0.0 + 4.70 = 9.11.4.7.0.$$

The ends of baktuns or katuns in the Long Count were often
recorded by the respective glyphs with the required coefficients in addition to a prefix or prefatory glyph which often
included a hand sign. This type of expression usually followed a Calendar Round date and anchored that date in the
Long Count. Examples of such period ending compounds are
illustrated in figure 11.26, a-c. The first describes the
joining together of 14 katuns on 6 Ahau 13 Muan, an event
which can only take place at Long Count position 9.14.0.0.0.
The second example records a seating of the tun on 12 Ahau 8
Ceh and notes that it was also the joining together of 11
katuns. This information securely anchors the given date
in the Long Count chronology at 9.11.0.0.0. The last example
gives an abbreviated formula for the base date of Maya chronology which can be glossed as "4 Ahau 8 Cumku, at the joining
together of 13 baktuns". It was seen earlier that the Initial Series formula for this zero date is 13.0.0.0.0 and it
is clear that the briefer version recalls the initial gathering of 13 baktuns in that formula.

A period ending compound signifying the joining together
or end of a tun and employing the same tun variant found in
the "seating of the tun" glyph is of frequent occurrence in
the inscriptions. It consists of the tun variant surmounting
a hand and often takes a prefix chosen from the same group
prefixes found with the Distance Number Introductory Glyph.
Even though the compound is written without numerical coefficients, the information that a particular Calendar Round date
occupies a tun ending position is, in practice, sufficient to

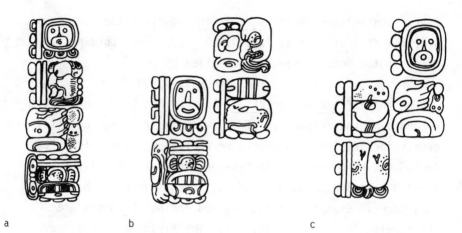

a   b   c

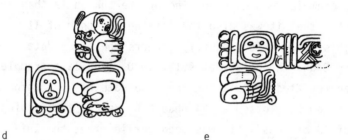

d   e

Fig. 11.26. Period ending compounds: (a) Piedras Negras, Stela 3, F7-F10;
(b) Palenque, Temple of the Inscriptions, Middle Panel, B1-A3; (c) Palenque,
Tablet of the Cross, D3-C5; (d) Palenque, Temple of the Inscriptions, East Panel,
L5-L6; (e) Naranjo, Stela 29, H11-H12.

fix it in the Long Count. Indeed, there are 73 month positions on which a tun may end and 13 possible coefficients of
Ahau which means that a given combination of these events
will not recur until the lapse of 73 × 13 = 949 tuns. In
particular, the passage illustrated in figure 11.26d mentions
that a 5 Ahau 3 Ch'en is a tun ending and it can be shown
that its Long Count position is 9.8.0.0.0. The nearest
alternatives, 7.0.11.0.0 and 11.15.9.0.0, are much too early
and much too late for chronological positions in the historical era. Similarly, the passage in figure 11.26e states that
9 Ahau 13 Pop is the end of a tun and it can be shown that
its Long Count position must be 9.13.3.0.0.

Other chronological devices were used to record certain
tun counts within the katuns of the Long Count. The most
common references were to counts of 5 and 15 tuns, followed
by counts of 10 and 13 tuns. The notation for a count of 5
tuns consisted of a numerical coefficient of 5 attached to a
tun sign (usually the variant used in seatings of the tun and
endings of the tun) and having a u bracket as a prefix. The
example shown in figure 11.27a reads "6 Ahau 13 Kayab, 5
tuns" and can be fixed in the Long Count at 9.17.5.0.0.

Markers for 10 tuns and 13 tuns consisted of appending
numerical coefficients of 10 and 13, respectively, to one of
the tun variants. The example in figure 11.27b records "13
Ahau 18 Kankin, 10 tuns" while that in figure 11.27c records
"4 Ahau 8 Yaxkin, 13 tuns". The former can be fixed in the
Long Count at 9.10.10.0.0 and the latter can be fixed at
9.15.13.0.0. The prefix of the 13-tun glyph in the last case
is equivalent to the u bracket.

The 15-tun glyph is constructed in a different fashion.
It employed a coefficient of 5 with a tun variant and an
additional prefix as illustrated in figure 11.27d. The

a

b

c

d

Fig. 11.27. Tun counts within the katun: (a) Quirigua, Stela C, D11b-D12; (b) Palenque, Tablet of the Sun, Q14b-Q15; (c) Tikal, Stela 5, C3-C4; (d) Piedras Negras, Lintel 2, X11-X12.

compound signified that 5 tuns were lacking in the count of
the katun. Thus, the present passage reads "4 Ahau 13 Mol, 5
tuns lacking". It can be fixed in the Long Count at
9.11.15.0.0. The given date is 5 tuns short of the katun
ending at 9.12.0.0.0.

NOTES

1. In this regard, a standard source is Maya Hieroglyphic
Writing: An Introduction by J. Eric S. Thompson (1971).
This work is a compendium of glyphic knowledge as it stood in
1950 when the book was first published. In the years since,
considerable progress has been made in the study of the Maya
script. An excellent account of the period of ferment from
1950 to 1967, a period which saw the development of new
approaches to old problems, has been written by David H.
Kelley (1976).

In the present study, citations are only provided for
results which draw on developments subsequent to Thompson's
publication. The illustrations are by the author unless
otherwise stated. In most cases they have been traced or
redrawn from earlier sources which are too numerous to
mention.

2. The kin glyph in figure 11.4a is usually restricted to
descending series but on Altar 1 from Naranjo it occurs in an
ascending series. The glyph in figure 11.4k is usually
restricted to an ascending series but occurs in a descending
series on a monument from Xcalumkin (Closs 1979).

3. Thompson has argued that this and other zero glyphs
have the connotation of completion. A good critique of this
interpretation has been made by Cesar Lizardi Ramos (1962).

4. The computational approach followed here is due to
Lounsbury (1978, pp.769-771).

5. The historical data in the monumental inscriptions was first revealed by Tatiana Proskouriakoff (1960, 1963, 1964). This particular text has also been examined by Closs (1982).

6. The first paper treating of the dynastic records at Palenque was published by Heinrich Berlin (1968). This was followed by a more comprehensive paper of Peter Mathews and Linda Schele (1973). The ruler in the text under consideration predated the sequence of rulers discussed by Mathews and Schele and is not mentioned in their paper. However, given the structure of the dynastic records which they elucidated it is a simple exercise to derive the information presented.

7. An analysis of the complete text of the sarcophagus lid has been published by Lounsbury (1974).

8. The precise mechanism used to achieve this has been closely examined by Closs (1977a). I have proposed a solution which differs somewhat from the conventional interpretations described by Thompson (1972, pp.62-70) and Lounsbury (1978, pp.776-789), but which I believe is more in keeping with the given data and the probable usage of the Dresden Codex.

9. There is no doubt as to the specific numerical value of the various heads. However, there are problems with some of Thompson's identifications of these heads with specific deities. With the exception of the head for 5, I have followed Thompson's list but have queried some of these as uncertain. Kelley (1976, p.96) regards the deity of 6 as an unknown goddess and considers that the deity of 11 is probably feminine. The deity of 9 is almost certainly more of a jaguar god than a serpent god.

10. These zero variants have been noted in publications of Sylvanus G. Morley (1915, pp.199-201), J.A. Graham (1972, pp.34-37) and Closs (1978).

ACKNOWLEDGEMENT

This work has been supported by a research grant from the Social Sciences and Humanities Research Council of Canada (410-79-0448).

12. The Zero in the Mayan Numerical Notation

A. Seidenberg

Similarities in the cultures of the Old and New Worlds
were already observed at an early date, that is, from early
post-Columbian times, and the problem of explaining the similarities has tried the intellect of successive generations.
The first solution was theological, explaining the similarities as due to the Great Deceptor, Satan. Beyond that, the
explanations put forward have had two poles, the Diffusionist
and the Independent Inventionist; or, one could also say, the
historical and the psychological. According to the Diffusionist theory, various widespread practices and beliefs are
not the spontaneous reactions of the human mind to environing
conditions but are the product of special circumstances:
having arisen at some spot, they spread, that is, diffuse.
The diffusionist tends to regard the duplicate appearance of
a cultural element, say an invention, as evidence of historical connection. Correspondingly, the independent inventionist tends to regard such recurrences as evidence that the
mind works similarly under similar circumstances.

The poles are, of course, but points in lines of argument
and require amplification and qualification. For example,
the theory of diffusion does not say that all culture has
arisen at one spot, but only that each of its elements has.
On the other hand, the Independent Inventionist is ready to
allow some exceptions: unless one holds that man is autochthonous to America, or holds some similar notion, one is
bound to allow that some of the culture of the American Indians comes from the Old World.

Thus, A.L. Kroeber held that America was peopled about
10,000 years ago by Asians coming in over the Bering Strait.
This would account for the mesolithic culture of the American
Indians; however, he would not allow that any higher cultural
element, except for some minor spill-overs from Northeast
Asia to Northwest America, came from the Old World: all was
developed here by the American Indians themselves. Not a
single element of pre-Columbian culture beyond the mesolithic
in, say, South America, or even Middle America, is, according
to Kroeber, due to contact with the Old World.

In the first edition of his book Anthropology, as well as
in the expanded version of 1948, Kroeber (1923) speaks of the
culture areas of America: these are some 14 areas into which
anthropologists have divided America, from the Arctic, Northwest, California, Plateau, Mackenzie-Yukon, Plains, Northeast, Southeast, and Southwest areas in the north, to Mexico,
Columbia, and Andean areas in the middle, and the Tropical
Forest and Patagonia in the south. In the first edition,
Kroeber (1923, fig. 35) has a nice "diagram illustrating the
occurrence of some representative elements of culture in the
various areas of America". He considers that the elements
common to all the culture areas (rows 1 and 2 of the diagram)
were brought in by the Asian immigrants. These include:
Dog, Bow, Harpoon, Firedrill, Woven and Twined Basketry,
Family Groups, Men's House, Shamanism, Crisis Ceremonies
(especially for Girls at Puberty), and Whipping of Boys. The
Spear-Thrower, which is lacking in some areas but is thought
to have existed in them once, is also included. Of higher
elements Kroeber considers (amongst others): the Magic
Flight (a folkore motif), Flood Legends, the Double-headed
Eagle, Pan's Pipes, Proverbs, Bronze, Exogamy, the Zodiac,
the Zero, and, in the 1948 edition, a Pachisi-like game

called Patolli. With the exception of the Magic Flight,
which has a rather limited distribution in northernmost America, and the proverb, which according to Kroeber did not
occur in pre-Columbian America,[1] all have developed independently in the New World. It would be useful to go over all
of Kroeber's considerations, in order better to understand
his thought and in order to evaluate the Independent Inventionist position, but here we must confine ourselves to the
Zero. It is noteworthy that with the Zero Kroeber considered
himself on especially firm, indeed absolutely firm, ground.

By the Zero we refer to the use of a symbol for zero in a
numerical position notation, known to us in our familiar decimal notation, but also known to the Maya in a 20-system,
that is, a system based on 20 as ours is on 10. But before
considering Kroeber's remarks, we will recall the history of
the zero and of positional notation in the Old World. We
will give this history as known now: it contains one essential point not known in 1923 (though known by 1933).

Already at the time civilization comes onto the scene in
written documents, Egypt and Sumer could count to some high
numbers and had a written notation for them. The Egyptians
had signs for one, ten, hundred, thousand, ten thousand, and
a hundred thousand, and wrote other numbers by juxtaposing
these signs. The system does not proceed to infinity, but is
regular as far as it goes. Similarly, the Sumerians had
signs for one, ten, sixty, and sixty times sixty. The Egyptian signs for one and ten are rather different:

one = $|$,   ten = $\cap$.

In Sumer the signs for one and sixty are the same except
that the sign for sixty is bigger: clearly 60 is regarded as
a "big 1". This is the first step towards a positional 60-
system: in such a system 1 and 60 must be written with the

same sign; and, further, the signs must be of the same size. By 1800 B.C., by which time the Old-Babylonians had taken over, the 60-system had developed into a full and regular positional notation based on 60. This notation had some defects, for example, it lacked a sign for zero, but even so it was an efficient device: the Old-Babylonians had a magnificent mathematics and made elaborate calculations (van der Waerden 1961, Chs. 1-2; Neugebauer 1957, Ch.1).

In a 10-system, our system, 21 stands for $(2 \times 10) + 1$ and 201 stands for $(2 \times 10^2) + 1$; similarly, in a 60-system 21 would stand for $(2 \times 60) + 1$ and 201 for $(2 \times 60^2) + 1$. If we lacked the zero, and wrote 201 as 21, we would not be able to tell the difference between the two numbers. The Old-Babylonians had precisely this dificulty; but they met it by leaving a little extra space between the symbols (Neugebauer 1957, pp.27-28). This, however, did not cure the trouble, as in a number like 2001, one would have to leave still more space. There is no question that the absence of a zero is a defect; and the Old-Babylonians did sometimes, though not often, make mistakes because of it.

By 300 B.C. (and perhaps much earlier) the (New-) Babylonians had a sign for zero (Neugebauer 1957, p. 27). It was not our "goose-egg", 0, but was taken over from a punctuation mark that looked a bit like our colon. This zero, however, was only used in internal positions: thus it would be used to distinguish 21 and 201, but not 21 and 210. This limitation on the use of the zero has often been called a defect, but it has its advantages, and amounts to the use of a "floating decimal point". When our students are taught logarithms, they must unlearn the "fixed" decimal point. Or even in elementary arithmetic, the student is taught that to divide by 2, say, one may multiply by 5 and adjust the

decimal point. More generally, division by any number ($\neq 0$)
becomes a multiplication by the reciprocal number. The Old-
Babylonians fully recognized this aspect of their notation.
If computations were confined to additions and subtractions
of whole numbers, the lack of the zero at the end might prop-
erly be called a defect; but for multiplication (and division), which is the principal computational application of
the system, the "floating decimal point" is very much in
order. Thus those who are quick to note the "defect" of the
New-Babylonian use of the zero, or see ethnological implications in it, are perhaps a bit off the mark.[2]

The Greeks of classical times did not have a positional
notation, but still they were able to compute fairly well,
provided the numbers weren't too large (van der Waerden 1961,
pp.45-50). But, later, when it came to astronomical computations, they were glad to have the Babylonian 60-system available. The sexagesimal notation was, however, rarely applied
consistently by the Greeks; thus Ptolemy (c. A.D. 150) even
in the Almagest writes the fractional part of the number
sexagesimally, but the whole part in the older Greek style
(van der Waerden 1961, p.50; Neugebauer 1957, pp.16, 22).
This was not serious, though, as the whole numbers were not
large, and the fractional parts contained the information
that could not be transcribed easily.

The Greeks of Ptolemy's time did not have our sign for
zero; rather they had a sign $\overset{\text{o—o}}{\text{o}}$, which varied, in varying
manuscripts, to forms like $\overline{\text{o}}$ or $\ulcorner\text{o}\urcorner$ or $\lceil\text{o}\rceil$. Thus the sign
does look a bit like our 0 with embellishments. The bare
0-like form does, indeed, appear in the works of Ptolemy, but
the extant Byzantine manuscripts of these works date from a
later time. Still later the Arabs continued to write $\overline{\text{O}}$ (Neugebauer 1957, p.14).

Other contributions were made by the Hindus. Our signs 1,
2, 3, 4, 5, 6, 7, 8, and 9 come from them, though not quite
in these forms. These symbols go back to the third century
B.C., though not as part of a positional system; indeed, the
signs for 10, 20, 30, et alia are also connected figures, and
the sign for 60, for example, is not like that for 6. (In
our decimal system, the signs for 10, 11, 12, etc., are disconnected figures; this results from the place value principle.) Eventually a zero sign was added and a positional
system was employed. The epigraphical evidence for this is
scanty: on a proclamation of a gift in A.D. 595, the corresponding Hindu year 346 is written with the signs for 3, 4,
and 6. This evidence, if authentic, shows a positional 10-
system in use in India before A.D. 600. The earliest occurrence of a zero sign in India is, according to Datta and
Singh (1962, p.41, no. 4, see also nos. 7, 19 and 20), seen
in an inscription of the eighth century, where 30 is written
in decimal place value notation (van der Waerden 1961,
pp.51-54).

There is some outside testimony for the decimal positional
notation in India. In a work of 662 Severus Sēbōkht, a Syr-
ian bishop, extols the science of the Hindus and speaks of
"their valuable method of computation ... I wish only to say
that this computation is done by means of nine signs". Presumably he should have said ten, not nine. Al-Khwarizmi (fl.
c. 825) was more precise. He wrote: "When (in subtraction)
nothing is left over, then write the little circle, so that
the place does not remain empty. The little circle has to
occupy the position, because otherwise there will be fewer
places, so that the second might be mistaken for the first"
(Cajori 1919, p.461; van der Waerden 1961, p.58).

Even more unassailable evidence comes from the practice of

Hindu astronomers in memorizing whole mathematical tables.
To do this they versified the numbers: in place of 1 they
said šaši, 'moon', because there is only one moon; for 2 they
said "eyes", "arms", or "wings"; etc. Zeros were also mentioned; the number 1021 might be expressed as follows:

šaši - paksa - kha - eka

moon - wings - hole - one

Note that the order here is the reverse of ours, and of the
Greeks. This versifying was the practice around A.D. 500.

Just before this time lived the astronomer and computer
Āryabhata, who, according to his own statement, reached the
age of 33 in 499. He had a different, and non-positional,
way of versifying numbers. In this method, the 25 consonants
from k to m (in Aryabhata's alphabet) have the values 1 to 25
(e.g., c = 6, g = 3, ñ = 5, ch = 7) and the other consonants,
from y to h, have the values 30, 40, ..., 100. These can be
combined with the vowels, a, i, u, r, etc., standing for
powers of 100. Thus the number 57 75 33 36 was versified
as:

cayagiyiñušuchlr

6 3 3 3 5 7 7 5 .

A zero is not needed in this system. Bhaskara I, a pupil of
Āryabhata, used an improved system of the kind mentioned,
which is positional and has a zero. Later, the order was
reversed. These facts suggest that the positional system was
introduced by the astronomers and computers around 500 (van
der Waerden 1961, pp.54-55).

It would appear, then, that around A.D. 500 whole numbers
were written in a decimal position system, though the astronomers continued to write the fractions sexagesimally. Thus,
we get our way of writing whole numbers in the 10-system from
India.

The sign for zero in Byzantium was a little circle and the sign for zero in India was a little circle. Is this a case of independent invention?

Neugebauer (1957, p.189) expresses the following opinion: "It seems to me rather plausible to explain the decimal place value notation as a modification of the sexagesimal place value notation with which the Hindus became familiar through Hellenistic astronomy".

Van der Waerden (1961, p.56), following Freudenthal, sums up his opinions as follows: "Before becoming subject to the Greek influence, the Hindus had a versified positional system [sic, but presumably he meant non-positional, as with Āryabhata], arranged decimally and starting with the lowest units. They had the digits 1-9 and similar symbols for 10, 20, ... Along with Greek astronomy, the Hindus became acquainted with the sexagesimal system and the zero. They amalgamated this position system with their own: to their own Brahmin digits 1-9, they adjoined the Greek 0 and they adopted the Greek-Babylonian order".

To complete the account of our decimal system, we add the following remarks. After the Hindus, the next step was taken by Al-Khāshī, the astronomer royal in Samarkand. Shortly before 1429, he had invented the decimal analog of sexagesimal fractions (Neugebauer 1957, p.23; Smith 1923-1925 (1), p.290), writing 2π somewhat as follows:

                    Integer

        6   2831853071795865   .

The final touch is the "decimal point". Pellas used the decimal point already in 1492, but the first to use it operationally was Christoff Rudolph in 1530. However, it was long for this invention to come into general use. Even Simon Stevin, who wrote a very influential pamphlet on the decimal

system, published in 1585, did not have it. The decimal
point came into general use in the 17th century, in connection with logarithms (Smith 1923-1925 (2), pp.238-245). Note
that the development of our decimal system took some 4500
years.

Except for the "decimal point", the (New-) Babylonians had
a complete place-value notation based on 60 and used it consistently. The Greek astronomers began the inconsistency
(still with us, as in 132°5'7") of indicating the integral
part of a number in a 10-system, and the Hindus continued it.
Thus the Greeks had no occasion, at least in the tables of
the Almagest, to use the zero at the end of a whole number,
though they used it at the beginning of a fractional number,
as did the New-Babylonians (Neugebauer 1957, pp.10, 20).
What the Babylonian really lacked was a "decimal point", so
he had to keep the absolute magnitude of the number in mind
(as we do when we use the slide rule). In the case of whole
numbers this can be done by the use of zeros at the end, and
this is an advantage of the Hindu notation for numbers
restricted to be whole. It should be kept in mind, though,
that the sexagesimal place value notation was a mathematical
device, not a method for denoting the small whole numbers of
daily life.

Let us now look at what Kroeber says about the Zero. He
begins by informing us that "our zero, along with the other
nine digits, appears to be an invention of the Hindus approximately twelve or fifteen hundred years ago". Now in 1923,
when he wrote this, he did not know of the New-Babylonian use
of the zero, but it had long been known, and presumably he
knew, that the Old-Babylonians already had a "position value"
notation. His silence on this point is strange, unless he
meant to narrow the issue down to the zero alone; but this

does not appear to be the case, partly from what he says explicitly and partly because he will claim that the Maya not only invented the zero but also everything else about their arithmetical notation.

Kroeber then reminds us how useful our system is for multiplying numbers, and how cumbersome the arithmetic operations would be with Roman numerals. But what he does not tell us is that there is no evidence -- at least he gives us none -- showing that the Maya ever used their notation for multiplying numbers, or even that they ever multiplied two numbers, with or without their notation.[3] We can agree that the Maya handled some large numbers and that they could add and subtract them. We do not wish to minimize what can be done with addition and subtraction, but still that's not to say that the Maya could multiply. Of course, here it is partly a matter of definition, and a Maya priest could have multiplied 23457 by 432, say, by repeated additions of 23457. In additions the notation can come in, but we still do not know whether the notation was used in finding a sum; conceivably the sum was worked out separately, say on an abacus, and the answer merely registered in their notation.

The Chinese, Kroeber tells us, had signs for "ten times", "hundred times", and so on; let us say T, H, etc. Then they could represent 1888 as 1Th 8H 8T 8, and 1005 as 1Th 5. This fails to be a place value notation, but two steps would make it so: introduce the zero and drop the T, H, Th, etc.

Before proceeding with Kroeber's remarks, it may be well to describe briefly the Maya counting. In their vocal language they had a regular vigesimal system, with non-composite words for 1, 20, 20^2 (= 400), 20^3, 20^4, 20^5, and 20^6. Their vigesimal notation, it has been said, was "usually" applied to counting time. Now the Maya had various periods of time:

first, the day; then a period of 20 days, a uinal, which we
will refer to as a "month"; 18 of these made a tun, a period
of 360 days. A sequence of numbers, found on the monuments
or in the codices, say 9.16.4.10.8 (as it is usually written
in the literature), gives the amount of time that has passed
from the zero date. In the example 9.16.4.10.8, the 8 refers
to 8 days, the 10 gives a number of "months", the 4 gives a
number of tuns, the 16 refers to 16 periods of 20 tuns, and
the 9 to 9 periods of 20^2 tuns (or 144000 days). Thus a unit
in any position is equivalent to 20 units in the next position to the right, except that a unit in the third position
from the right represents only 18 of the next smaller unit.
This system has been called a "modified vigesimal system".
The reason for this is that the author of the phrase, H.J.
Spinden, considered the inscriptions to give a day count, as
did S.G. Morley, another worker in the field (see Closs
1977b, p.18). It's hard, for some of us anyway, to believe
this. J.E. Teeple (1931, p.35) believed, rather, that the
last two digits give a day count, up to 360, and that the
others give a tun count. Thus the inscriptions give a tun
count plus a day count (up to 360); or even more precisely,
though we need not fuss over this point, they give a tun
count plus a "month" count plus a day count (Closs 1977b).
That there should be such a confusion over a trifle amongst
the principal workers indicates how unlikely it is that they
came upon vigesimal computations in their researches.

   Teeple (1931, p.36) says: "Strictly speaking the numerals on the monuments do not have position value, because they
nearly all have symbols for the tun, katun, baktun, etc.,
just as we might use symbols for the terms ten, hundred,
thousand, etc. ...". Except for the zero, this is precisely
what the Chinese did. There is one early date, according to

Teeple, using a pure position value system, and Michael D.
Coe (1976, pp.112-113) has given others. Moreover, in the
Dresden Codex, which dates from a much later time, nearly all
numerals are to be read from position value only. The Maya,
just like the Chinese, wrote their numerals in columns with
the smaller units below the larger.

Speaking of the Mayan zero, Kroeber writes: "This Maya
development constitutes an indubitable parallel with the
Hindu one. So far as the involved logical principle is
concerned, the two inventions are identical. But again the
concrete expressions of the principle are dissimilar. The
Maya zero does not in the least have the form of our or the
Hindus' zero. Also, the Maya notation was vigesimal where
ours is decimal. They worked with twenty fundamental digits
instead of ten ... Obviously there can be no question of a
common origin for such a system and ours. They share an idea
or a method, nothing more. As a matter of fact, these two
notational systems, like all others, were preceded by numeral
word counts. Our decimal word count is based on operations
with the fingers, that of the Maya on operations with the
fingers and toes. Twenty became their first higher unit
because twenty finished a person."

We can agree that the inventor of the 20-counting had ten
fingers and ten toes and that the Maya also had ten fingers
and ten toes, but it still does not follow that the Maya
invented their 20-counting. The vigesimal system of counting
has a widespread distribution; examples of it occur in all
the continents and in Oceania. The Sumerian number system
which is usually called sexagesimal and sometimes decimal-
sexagesimal is, rather, vigesimal-sexagesimal, especially in
the count up to 60 (Seidenberg 1965). Thus, 20-counting was
in existence some 3500 years before we see it in operation

with the Maya. There is no reason to think the Maya invented
this system, and plenty to think they didn't.

As to the form of the sign for zero: in the Old-World we
see the Greeks taking over the sexagesimal system, along with
the zero, from the (New-) Babylonians; but they wrote the
"digits" from 1 to 59 in their familiar way, and as for the
zero, they invented or took over some familiar sign for that,
too.[4] It is thus easily conceivable that the Maya, too,
learned of the zero from others, but used some local sign for
writing it.

For Kroeber the place value principle is a mere "idea or a
method, nothing more". Why shouldn't anybody be able to
think it up? Yet in the Old-World things were not so simple.
The position value principle had been around for some 2000
years before the Hindus learned of it. They then combined
the place value principle, which they learned of along with
sexagesimal numbers, with their 10-counting to produce a
decimal position number notation. Here was something new,
for it wasn't there before. If one wishes, one can say that
they invented something, but what? It's better not to call
what they did an invention at all; rather, it is an application of a borrowed idea to a local situation, hardly even a
modification of the idea, yielding, it is true, an improvement on the local system but not necessarily an improvement
on the source. In the case of the Maya it is thus easily
conceivable that they learned of the place value principle
from 10-counters but adapted it to their familiar 20-count-
ing, just as the Hindus learned of the principle in 60-form
but adapted it to their 10-counting. There is no more reason
to think of the embodiment of the place value principle in a
20-system as an independent invention than to think of its
embodiment in a 10-system as an independent invention.

We have seen that the Chinese and the Maya wrote their
numbers in a somewhat similar way, though to different bases.
Thus, every aspect of the Mayan notation, except the zero
itself, could be matched with things known already in 1923,
and probably known to Kroeber, to have existed in the Old-
World. The zero is but part of a complex. Even if we were
to concede the zero to be an independent Mayan invention, it
would appear as a one-step affair; and, moreover, one that
led to nothing. For all we know, there is nothing the Maya
did with their zero, except for decorating monuments, that
they could not equally easily have done without.

In 1923 Kroeber continued: "It is interesting that of the
two inventions of zero, the Maya one was the earlier. The
arithmetical and calendrical system of which it formed a part
was developed and in use by the time of the birth of Christ.
It may be older; it certainly required time to develop. The
Hindus may have possessed the prototypes of our numerals as
early as the second century after Christ, but as yet without
the zero, which was added during the sixth or according to
some authorities not until the ninth century. This priority
of the Maya must weaken the arguments sometimes advanced that
the ancient Americans derived their religion, zodiac, art, or
writing from Asia. If the zero was their own product, why
not the remainder of their progress also? The only recourse
left the naive migrationist would be to turn the tables and
explain Egyptian and Babylonian civilization as due to a Maya
invasion from Yucatan."

We can agree that it takes time for things to develop, but
the view that the Mayan arithmetical notation developed in
Yucatan is admittedly based on no evidence. In the Old World
we can pretty well see the development of arithmetical notation from scratch. In Yucatan, as elsewhere in Mesoamerica,

we have only the finished product.[5]

So much for what Kroeber wrote in 1923 on the zero. By 1948 he had learned of the New-Babylonian zero. In the revised edition of his Anthropology he first copies what he had in 1923, with some minor changes, and then he launches into the new evidence. "It has only recently become clear," he writes, "that the zero as a position numeral was invented a third time, and that the earliest of all. This was in Mesopotamia, among the Semitic Neo-Babylonians, centuries before either Hindus or Mayas." What had made him so confident, even cocky, in 1923 was the chronology, the fact, or what was supposed to be a fact, of the priority of the zero in America. In 1948 he is obliged to retract, but nonetheless he blandly repeats the priority argument of 1923. He insists that we are dealing with a third case: "The idiosyncrasy of this system [with its higher unit of 60] is enough to assure the separateness of the Mesopotamian invention." As we have seen, this was not the opinion of Neugebauer, Freudenthal, or van der Waerden.

Among the minor changes of 1948, Kroeber deletes the barb against the "naive migrationists" from the section on the zero. He would have done better to have removed the section itself.

The Independent Inventionists can regard the Zero with scant satisfaction.

NOTES

1. See, however, Loeb (1964).

2. For an error resulting from a "misplaced" decimal point, see Neugebauer (1957, p.28).

3. Kroeber (1948, p.469) does in his later edition, although a couple of pages earlier he repeated the remark

about the cumbersomeness of the Roman numerals.

4. It has been suggested that the Greek O is an omicron, an abbreviation of their word ονδεν for "nothing". Neugebauer (1957, p.14) considers this implausible because the omicron already represented a numerical value, namely, 70. Van der Waerden (1961, p.50) counters this by noting that there would never be the occasion to use O for 70 in a sexagesimal expression: thus for the Greeks the O would always stand, without confusion, for zero in the fractional part of a number (and would stand, also without confusion, as 70 in the integral part). Still, Neugebauer makes the point that the manuscripts of an early date -- the first three centuries B.C. -- do not support this, in that the bare O-form does not occur; only forms like Ō or related forms occur. As a way of reconciling these views, I would suggest that the O does come from ονδεν, but since the omicron already had a numerical significance, it was felt necessary to embellish it.

5. The oldest known dated monument in Mesoamerica, according to Coe (1976), is Stela 2 from Chiapa de Corzo, Chiapas, Mexico. In the Thompson correlation of the Maya and Christian calendars, this dates from 37 B.C. In the Maya notation it is shown as 7.16.3.2.13 and already employs the familiar "bar and dot" numerals as well as the place value principle.

13. In Search of Mesoamerican Geometry

Francine Vinette

Determining the mathematical knowledge in Mesoamerica has
proved an enormous challenge to researchers. The task is
made particularly difficult because of the paucity of Pre-
Columbian written materials and because texts written by
early Spanish inquirers provide little specific information
in this area. The researcher, therefore, is forced to rely
on ingenious conjecture and on verifiable methodologies in
order to derive the mathematical knowledge of these fascinating civilizations.

This paper presents the work of a number of researchers,
to support the hypothesis that knowledge of geometrical concepts was widespread in Mesoamerica. The paper does not
attempt to deduce the specificity of geometrical knowledge
but rather, reports and assesses the literature pertinent to
the area, with the intention of provoking discussions that
may lead to further studies and developments in this relatively unexplored field.

ARTIFACTS AND MONUMENTS

Although few written documents of the ancient settlers of
the Americas have survived, Mesoamerica abounds in artifacts
that are in fact real masterpieces. Many of their works of
art, such as painting and sculptures, are perceived, by contemporary scholars, to be the products of talented artists.
However, the final execution of these works often exhibit a
lack of precision and gives rise to the learned opinions of
Miller (1973), Sanders (1977), and Robertson (1977; 1983),
that a master laid down the designs leaving journeymen or

apprentices to complete the works. Evidence of the use of
templates and patterns, in the painting and sculpture techniques of Mesoamerica, are corroborated by the same three
authors. They further point out that the planning of space
in murals, stelae, and carved reliefs, for paintings and
sculptures, as the case may be, had been arranged before the
outlines of the work of art were actually drawn.

For example, the measures of the profile figures of Room 2
at Tepantitla (Miller 1973, p.33), show that the painting
space was measured, "otherwise we cannot account for the fact
that the figures fit so well into the allotted space, that
they are roughly the same size and that they have similar
spacing between them." The mural circles of Substructure 3
of Zone 2 is also worth mentioning because it not only shows
an intention of measurement of the painting space but provides evidence for the use of a pointed instrument described
by Miller as a "kind of farmer's compass". The mural is made
up of 15 circles that were first marked with this instrument
and then painted in red (not always respecting the engraved
contour). The first 13 circles are 32 cm in diameter and are
separated by almost 7.5 cm. The last two engraved circles
are 33.5 cm and 34.5 cm in diameter, which once painted
became 35 cm and 38 cm, respectively, and the spacing between
them is 11.5 cm. The sudden differences in diameter and
spacing of the last two circles testify to a certain measured
design but also indicate a lack of precision in their execution.

With respect to the existence of the compass in Mesoamerica, it may be noted that it is included in the inventory of
Aztec construction tools listed by Guerra (1969, p.43). In
fact, he not only mentions the compass (tlayolloanaloni), but
also the plumb (temetzetepilolli), the level (quamniztli),

the square (tlanacazanimi), the trowel (teneztlasoloni), and
the wedge (tlatlilli).

In her study of the murals of Coba, Fettweis-Vienot (1980)
reveals, not only that the Maya had planned the space of the
mural, but also found evidence that there was, in the composition, a definite intention of symmetry. This intention, as
she points out, so evident in Maya writing, could very well
have been manifested in another artistic expression. She has
noted that there is a drop of blue paint found on the capital
of a column at Coba, which divides it exactly into two parts
measuring 83.4 cm each. She also relates other instances of
a perceptible intention of binary division: Structure III-B
of the Grupo del Rey in Cancun, the interior doorway of the
Castillo at Tulum, Structure 64 at Tancah, Structure B at San
Miguel de Ruz, as well as instances in Playa del Carmen,
Paalmul, and Cacaxtla, each of these structures having its
middle point clearly marked by a dash or a square (Fettweis
1973; Fettweis-Vienot 1980). The clues which mark intended
symmetries lead to "better understanding of the organization
of the composition, and of eventual repetitions of elements
on both sides of a central axis" (translated from Fettweis-
Vienot 1980, pp.27-28). Proportions, colors, and the relative position of each element are given special attention in
her analysis. But, it seems to me from her work, as well as
from the studies of Miller on murals 1 and 4 of Portico 11 at
Tetitla, and by Rodriguez (1969) on the murals of Teotihuacan, that any analysis of artifacts, for the purpose of
determining Mesoamerican knowledge of geometry, must be com-
plemented by analyses of symbolism and iconography.

Evidence of geometrical knowledge may also be derived from
an examination of the composition of artifacts. In an analysis of the monuments of Tikal, Clancy (1977) proposed that a

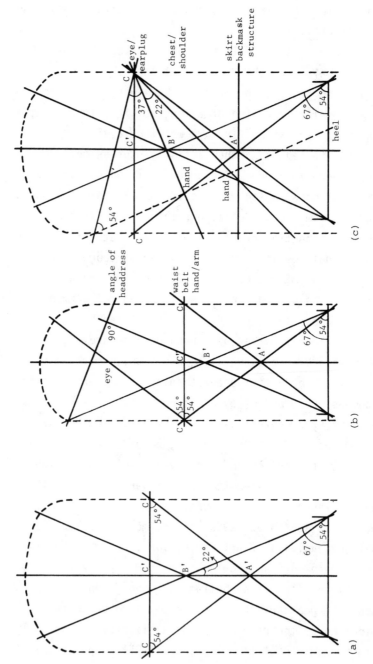

Fig. 13.1. (a) Compositional structure on Tikal stelae; (b) Compositional structure on Tikal stelae of the Middle Classic; (c) Compositional structure on Tikal stelae of the Late Classic (after Clancy 1977).

compositional structure, formed by the intersection and over-
lapping of isosceles and Pythagorean triangles, was pre-
established in them, implying therefore that the Maya had
access to the geometrical knowledge of these figures. From
the extremities of the lower horizontal basis of a stela,
Clancy drew two sets of diagonals of 54° and 67°. The points
of intersection A' and B' of these four diagonals, determine
the central vertical line. A horizontal line joins the end
point C of the two diagonals of 54°. These lines make up the
basic compositional structure illustrated in figure 13.1a.
Stelae of the Middle Classic exhibited a simpler organization
and composition when compared with the intricacy of the more
numerous relations of angles and triangles found in the compositional structure of the Late Classic stelae (figs. 13.1,
b-c). The critical points of Clancy's basic compositional
structure correspond to certain motif features. Table 13.1
summarizes the frequencies of these correspondences when
applied to Tikal stelae.

Although her criteria are not sufficient to infer conclusive results,[1] Clancy's analysis initiates an approach which
is open to further development and which may, in future, confirm her hypothesis of a prescribed compositional structure
for Maya stelae as well as the Maya cognizance and application of Pythagorean triangles (measures of 22°, 37°, 54°,
67°, 90° can be related to the angles of Pythagorean triangles).

Investigations of pre-established compositional structures
in monuments raise some crucial points which have to be taken
into account before asserting the Mesoamerican knowledge of
the geometrical concepts inherent in the compositional structure. The present state of the artifacts and monuments show
evident sign of erosion or fading or even, sometimes, of

Motif Feature	Stela										
	9	3	7	27	10	30	16	21	5	20	22
Belt Hanging	A'	A'			A'					A'?	
Apron Line			A'	A'							
Skirt Line						A'		A'	A'		A'
Backmask							A'	A'	A'		A'
Belt Line	B'	B'			B'						
Waist		B'		B'	B'	B'	B'				
Hand					B'		B'		B'		
Chest/Shoulder					B'		B'	B'	B'	B'	B'
Center Point of Stela	C'	B'	C'	B'	B'	B'	B'	-	-	-	-
Chest	C'			C'	C'						
Waist		C'	C'								
Earplug						C'	C'		C'	C'	C'
Hand	CC	CC									
Belt Line		CC	CC								
Chest Line				CC	CC						
Neck						CC	CC		CC		
Nose (Plug)										CC	CC
Top of Backmask									CC	CC	CC

Table 13.1. Compositional points and lines compared to motif features on Tikal stelae (after Clancy 1977).

destruction or alteration due to restoration. All such
factors reduce the precision of measurements and leave the
estimation of the margin of error to judgments, which, being
subjective, can be disputed. The major difficulty therefore
resides in the determination of whether a geometrical structure is being imposed on the artifacts rather than being
extracted from them.[2]

The unusual shapes of some constructions in Mesoamerica
have attracted particular attention of scholars, whose investigations and explanations of their peculiarities, are also
worthy of note for our purposes. For example, the astronomical function of the Caracol at Chichen-Itza (Aveni, Gibbs
and Hartung 1975) accounts for the peculiar position of
several of its architectural elements, which reflect both the
astronomical knowledge of the Maya as well as their ability
to construct precise right angles.

A second interesting construction is the arrowhead-shaped
Mound J at Monte Alban, imitated by Building O at Caballito
Blanco, which was interpreted as an astronomical observatory
by Aveni and Linsley (1972) and Aveni and Hartung (1981),
although its astronomical function has been contested by
Marcus (1983) and Spencer (1982), who consider the secular
function of these constructions to be more relevant. Spencer
(1982, pp.27-28) states: "Additional evidence for state
level decentralized decision making is found in the form of
an arrowhead-shaped building at the site of Caballito Blanco,
about 50 kms east of Monte Alban. This building is very
similar in form and orientation to Mound J at Monte Alban.
Mound J was an administrative facility serving a specialized
decision-making function possibly related to Zapotec military
activities". In addition to the astronomical and secular
function hypotheses for the resemblance of Building O and
Mound J, one could speculate whether the differences in the

orientations of the buildings are due to geomagnetical factors, while the resemblances are due to the practice of geomancy. These speculations are based on evidence of Carlson
(1976; 1977) suggesting possible geomagnetical knowledge and
geomantic practices in Mesoamerica, and stressing therefore,
the relationship between geometry and geomancy previously
alluded to by Hartung (1977).

The nunnery complex at Uxmal, also a peculiar construction, which shows a pair of right angles at the center of its
courtyard and parallel visual lines, was described by Aveni
(1982) and Aveni and Hartung (1982), to be a building built
on a hidden plan. These authors have also studied the parallelism of long versus short walls of buildings in Uxmal,
Chichen-Itza, and Palenque, and have noted a tendancy for the
long walls to be parallel while the short ones are not. This
peculiarity, as well as the previous examples from monuments
and artifacts, demand from us the acceptance that the Maya
did possess some system of geometry.

SITE PLANS

Geometrical relationships between monuments can be found
with relative facility through examination of site plans.
The real endeavor, in the context of a search for manifestations of knowledge of geometry, consists of identifying sufficient grounds to support the hypothesis that Mesoamericans
intended to display such relationships. The following examples from Hartung's (1977) analysis of the maps of Tikal and
Copan, are intended to underline his considerations concerning the non-coincidentality of the exhibited geometrical
forms.

In the map of Tikal, Hartung found an accurate east-west
baseline from the doorway of Temple I to the doorway of

Temple III (fig. 13.2). As he explained, equinoctial observations could have been made only before the construction of
Temple III or through (non existing) holes in the back wall
of one or both of the temples, these constructions having
different facade orientation and building time (9° E/N and
A.D. 700 for Temple I; 18° E/N and A.D. 810 for Temple III).
Even more relevant perhaps for the occurrence of the East-
West baseline, is the geometrical interrelationship of the
two temples demarcating this line, with respect to other
monuments in the site of Tikal. The East-West line forms
together with a South baseline from Temple III to Structure
5D-90 on the South side of the Plaza of the Seven Temples,
the equal sides of an isosceles right angle with right angle
at Temple III. Map measurements revealed an isosceles
triangle having Structure 5D-104 as vertex and this same line
as base (distances from Str. 5D-104 to Temple I and to Temple
III are 282 m and 283 m, respectively).[3]

Further geometrical concepts are also displayed by several
contemporary monuments of Tikal. Alignments from the doorway
of Temple I (A.D. 700) to the doorways of Temple IV (A.D.
750) and of Temple V (A.D. 700) form an almost perfect right
angle (89°57') at Temple I. Moreover, two North-South parallel lines connect the twin pyramids of Complex O (A.D. 731)
to Temple I and Temple V, and, incidentally, create another
right triangle at Temple I formed by the East-West line and
one of the pair of parallel lines, provided that these lines
have an exact North-South orientation. Another set of parallel lines connects the twin pyramids of Complex O with those
of Complex P (A.D. 751) and have approximately the same orientation as the axis of the North Acropolis.

Hartung's study of the map of Tikal shows the importance
of central doorways in the intent to justify the geometrical

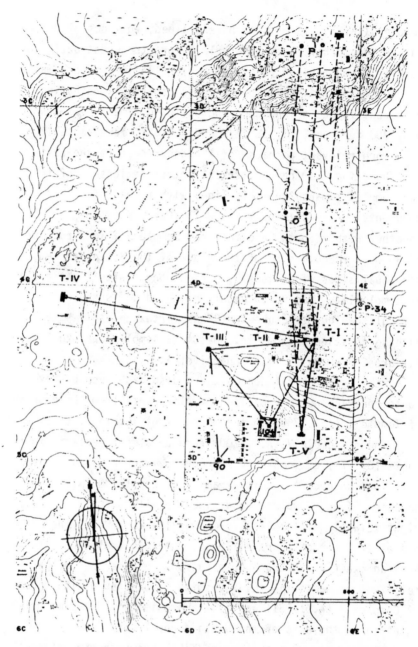

Fig. 13.2. Map of Tikal showing some geometrical relationships (from Hartung 1977, fig.9.1).

arrangement of monuments. By contrast, in his analysis of
the map of Copan, architectural masses such as altars,
stelae, platforms, or rock reliefs may have served as reference points in the setting out of the site (fig. 13.3).

The alignment of the platform in front of the entrance of
Temple 22 (and not its doorway) with the clear cut masonry
block in the stairway of Temple 11, determines a baseline
that is parallel to the visual line between stela M and Stela
O, as well as to the transverse axis of the ballcourt. Moreover, this masonry block in Temple 11 is the intersecting
point of three equidistant lines drawn from the central
marker in the ballcourt, the platform of Temple 22, and the
central of the three markers in the East Court of the Acropolis. This same architectural mass is also the vertex of an
isosceles triangle having for base the line between Stela M
and Stela O.

That the position of ballcourts should be related to other
monuments at the site, is not unusual in Mesoamerica. This
is the case in Tula and Chichen-Itza, where, in addition to
the resemblance between Temple B at Tula and the Temple of
the Warriors at Chichen-Itza, Aveni (1980) pointed out the
similar location of the principal ballcourts at both sites as
well as their identical orientation. Recent excavations at
Cerros revealed the existence of two Late Pre-Classic ballcourts having a general North-South orientation (Scarborough
et al. 1982). A broad North-South medial axis going through
the ballcourts and Structures 3 and 4 seems to bisect the
site and to intersect the westward primary axis of Structure
29B and its associated plaza at a point approximately equidistant from either ballcourt. The orientation to the west
of the central of three platforms at the summit of Structure
29B, while the other two platforms face the North and the

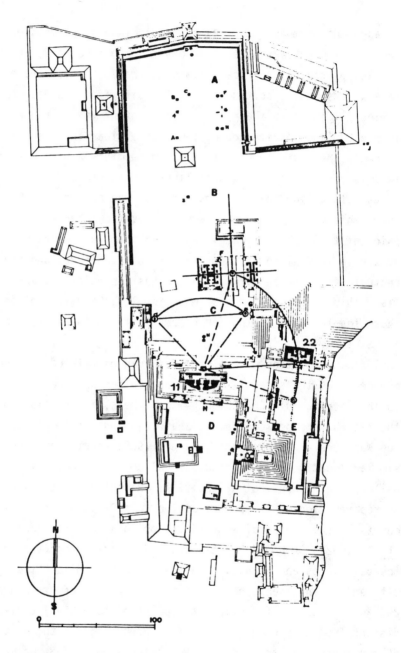

Fig. 13.3. Map of Copán showing some geometrical relationships (from Hartung 1977, fig.9.6).

South, support the hypothesis of a deliberate geometrical
relationship between these two ballcourts and Structure 29B.
The North-South orientation of these ballcourts reminds one
of the investigation that Taladoire (1979) undertook in order
to verify the hypothesis that ballcourts were aligned with
cardinal points to emphasize the proposed association of the
ball-game with the apparent movements of the sun. Because
the orientation of several ballcourts was still unknown, no
con- clusive results were obtained. Nevertheless, a symbolic
motivation for the orientation of ballcourts cannot be completely be ruled out.

Evidence that Mesoamericans observed the sky and possessed
an accurate knowledge of astronomy, has been assembled by
Aveni (1980) who, in his book, also updates several earlier
studies of archaeoastronomy. I now recall the astronomical
alignments found in Uxmal and Teotihuacan which, although
they have been described several times elsewhere, are particularly relevant in our search for geometrical concepts in
Mesoamerica.

The peculiar orientation of the Governor's Palace at Uxmal
(about 20° off the general axis of the site) has been noted
by Aveni (1975). A visual line, as viewed from the central
doorway of the Governor's Palace, perpendicular to its front
face joins the base of a half fallen stone column, the center
of a platform surmounted by a double headed jaguar, and the
ruins of Nohpat 6 kms away. This line also marks the southerly great extreme position of Venus rising in A.D. 750
(approximately, the building period of Uxmal). The evidence
of the importance of Venus in Maya cosmology and the accurate
observation of its movements, in addition to the Venus iconography exhibited in the monuments of Uxmal, argue strongly
for a deliberate astronomical alignment motivating the

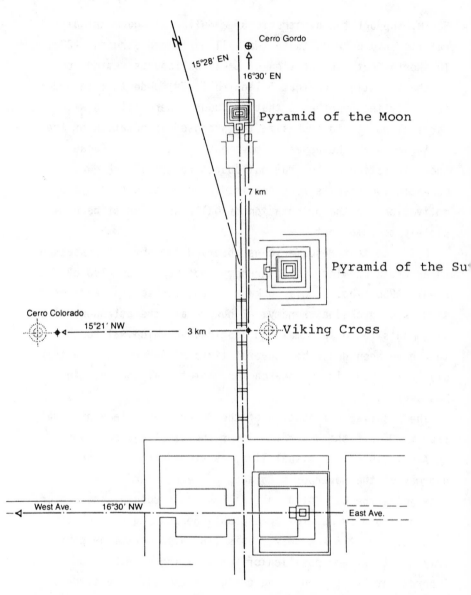

Fig. 13.4. The center of Teotihuacan and some orientations (from Aveni 1977, fig.1.2).

orientation of the construction. Consequently, the perfect
right angle displayed should not be considered coincidental.

If, at first sight, Mesoamerican site plans manifest a
noteworthy disorder in the arrangements of structures, Teotihuacan, the most imposing center in Mesoamerica, gives the
impression of having been constructed following a regular
grid (fig. 13.4). Indeed, the site seems to be built along
parallel axes to the main street, the Street of the Dead, as
well as along parallel axes to another principal street, the
East-West Avenue. However, measurements of this apparent
intended rectangular network show that these major streets,
the Street of the Dead and the East-West Avenue, do not cross
at a precise right angle but rather deviate from a 90° intersection by about 1°15'.

Aveni (1975) proposed an astronomical explanation for the
orientation of Teotihuacan. Having noticed the existence of
three pecked crosses, the Viking Cross (Teo 1) near the Pyramid of the Sun, one at Cerro Colorado (Teo 5), and the other
at Cerro Gordo (Teo 6), he discovered that the line Teo 1 -
Teo 5 is within 1° of an alignment with the Pleiades and
forms a very accurate right angle (89°53') with the Street of
the Dead, as well as the 90° crossing of the line Teo 1 - Teo
6 (16°30' E/N in Aveni 1977 and 17° E/N in Aveni 1980) with
the East-West Avenue.[4] The Pleiades' role in Mesoamerican
cosmology as well as Aveni's remark that "The Pleiades under-
went heliacal rising on the same day as the first of the two
annual passages of the Sun across the zenith, a day of great
importance in demarcating the seasons" (Aveni 1977, p.5),
could justify an intended right angle between Teo 1 - Teo 5
and the Street of the Dead. However, an astronomical motivation behind Teo 1 - Teo 6 and therefore the intentionality of
a perpendicular intersection of the East-West Avenue with Teo

1 - Teo 6, is not upheld, since Teo 1 - Teo 6 doesn't align with any significant astronomical body or event.[5] Consequently, an astronomical explanation for the off-set grid apparently based on the intersection of the Street of the Dead and the East-West Avenue, cannot be validated.

Assuming that the Teotihuacanos intended to build their site according to an orthogonal grid,[6] Chiu and Morrison (1980) proposed a hypothesis for the deviation from a true right angle for the crossing of the Street of the Dead and the East-West Avenue, as well as for the whole plan of Teotihuacan. They noticed that the sun on August 12 sets half-way behind Mound Xalpan at about 35 kms from their suggested location of Teo 6,[7] as well as behind Cerro Colorado at 285°6' viewed from the alignment Teo 1 - Teo 5,[8] an angle similar to the azimuth 286°25' of both the sun and the peak of Cerro Gordo. The builders of Teotihuacan could have constructed a perpendicular line or master line (Teo 1 - Teo 5) to the alignment with the sun from Teo 6, then draw a series of orthogonal lines to this master line. A second series of lines could then have been traced, but this time, perpendicular to the alignment Teo 1 - Teo 5. These lines should form an orthogonal grid, but as Chiu and Morrison explained, if the two identical astronomical alignments are not seen parallel in the ground, an offset grid will ensue. The elevation of the horizon as seen from Teo 1 - Teo 5 being 1°4' higher than from Teo 6 to Mound Xalpan lead to the apparent lack of parallelism of the alignments and caused therefore the deviation from a right angle that can be noticed in the site plan. The fact that the Teotihuacanos "could well hold the use of the celestial mark as a purpose in itself, more significant to them than the symmetry of a ninety degree crossing" (Chiu and Morrison 1980, p.S59), is an important observation.

Rather than an astronomical explanation, Drewitt (n.d.)
proposed a different argument to account for the offset site
plan of Teotihuacan. Drewitt constructed a grid having lines
parallel to the Street of the Dead and to the East-West Avenue, generally at intervals of 322 m or 400 units, one unit
of measure being suggested by Drewitt to be approximately
80.5 cms.[9] When superimposed on the site plan, Drewitt
noticed that this non-orthogonal grid follows all long
streets, many ceremonial structures and precints, and even
the deviation of the Rio San Juan by canals. In addition, he
enumerated other instances of a deviation from a 90° angle by
1°15' occurring in Mesoamerica. Such occurrences appear in
the relationships of Mound A and B at Kaminaljuyu, in the
Structures and complexes at Monte Alban, in the Central
Acropolis at Tikal, in the Structure B complex at Tula, and
in Complex A at La Venta. He concluded that this skewed
axial system "seems to have been a basic element in the layout of single buildings and of building complexes and in that
respect formed an integral part of construction practices"
(Drewitt n.d., p.9). Moreover, as Drewitt (personnal communication) pointed out, neither Teo 1 - Teo 5 nor Teo 1 - Teo
6 follow this network or any of the streets, construction
walls, or canals in the site plan. In addition, several
other pecked crosses have been found in Teotihuacan. It
would be worthwhile to check other astronomical alignments or
geometrical relationships in order to discuss the significance of their location on the site.

Finally, the hypothesis of Heyden (1975) based on the wor-
ship of cave and water deities, should be included in this
analysis of the planning of Teotihuacan. She believes that
the Teotihuacanos could have built the Pyramid of the Sun on
top of a sacred cave and then aligned the Street of the Dead

with the Cerro Gordo, their main source of water supply, and afterwards constructed the Pyramid of the Moon. In this interpretation of Teotihuacan's planning, the baseline towards Cerro Colorado is not mentioned. If the Street of the Dead was indeed aligned to Cerro Gordo, the astronomical orientation of the Cerro Colorado baseline towards the Pleiades (or August 12 sunset), would seem coincidental and the 90° angle between the Street of the Dead and this base-line, nonintentional.

In addition to the astronomical, geomagnetical, geomanti-cal, or purely functional considerations proposed in this section which might justify the manifestations of geometrical concepts displayed in Mesoamerican centers, effects of light and shadow have also proved to be worth investigation. In fact, the formation of isosceles triangles in the balustrade of the Castillo of Chichen-Itza on the equinoxes, the casting of Temple II's shadow exactly on the stairway of Temple I at Tikal, or the astronomical hierophanies in Palenque at sol-stices (Hartung 1977; Schele 1977), illustrate the astronomi-cal as well as the geometrical knowledge required in order for the arrangement of the structures or the construction of the structures themselves, to produce the desired visual effect. Since special astronomical or visual alignments can be observed on equinoxes or solstices (Aveni 1980), hieropha-nies on those important days in Mesoamerica are worthy of further investigation in the search for Mesoamerican geomet-rical knowledge.

SUMMARY

Occurrences of manifestations of geometrical concepts found on artifacts, monuments, and site plans have been pre-sented. The first part of this paper exhibited evidence that

space allotted for painting or sculpture was planned before
the Mesoamerican artist undertook the layout of his painted
or carved work. Analyses investigating the possibility of a
pre-established compositional structure for artifacts and
monuments, is a useful approach to the problem of uncovering
Mesoamerican knowledge of specific concepts of geometry, but
care must be taken that such geometrical knowledge is extracted from the artifacts and monuments rather than imposed upon
them.

Various hypotheses presented by researchers in diverse
disciplines have been surveyed and an attempt has been made
to determine the motivation behind Pre-Columbian site planning. Some of the arguments considered supplied justification for the intentionality of the geometrical concepts
exhibited in the site plans. In other cases the results are
debatable and require further development. Moreover, the
diversity of the considerations as well as the studies discussed in the first part of this paper, suggest that there is
an integration in Maya thought of astronomy, cosmology, art,
and site planning. There is little doubt that further investigations directed towards an evaluation of Mesoamerican
knowledge, will include geometry as a part of the amalgam of
Mesoamerican science and religion.

NOTES

1. Clancy's criteria are the following:

a) to be significant the angular and linear relationship
must be found to repeat on at least three different
monuments;

b) angular relationships were considered similar when
there was no more than a 2° variation between them;

c) a basic repeatable structure that would maintain the

above criteria must be common to at least 50 percent of the
monuments tested.

2. The compositional structure analyses of the Stucco
Palace of Acanceh (Hébert-Stevens 1972), the Castillo of
Chichen-Itza (Arochi 1977), and the Pyramid of the Sun at
Teotihuacan (Tomkins 1976, pp. 242-252) are examples where
the present state of the studied monument is overlooked and
where, at times, elements are forced to fit the author's
hypothesis. These authors then proclaim that the geometrical
concepts exhibited in the compositional structure were known
in Mesoamerica and deliberately applied in the construction
of the monument.

3. Exact measurements of the orientation of the South line
from Temple III to Structure 5D-90 and of the facade orientation of Structure 5D-90 and Structure 5D-104 as well as the
dates of construction of these structures might provide additional justification for the geometrical relationships displayed between these monuments of Tikal.

4. The relief and the poor visibility of the Pleiades at
this latitude at the time of construction of Teotihuacan may
account for the deviation of 1°.

5. Aveni (1977; 1980) has proposed Dubhue as a possible
astronomical alignment, but also noted that Dubhue is not an
astronomical body important in Mesoamerican star lore and
that similar alignments are not known in Mesoamerica.

6. Chiu and Morrison have proposed another way to construct a precise right angle that can be added to the other
possible methods for such construction described by Aveni and
Hartung (1982), methods that could have been easily used in
Mesoamerica. Let equidistant points A_i be marked on a segment AB by the use of long cords and from each one of these
points A_i, let arcs of fixed radius be drawn on each side of

AB to the right and to the left of each A_i. The intersec-
tions of these arcs, located between A_i's on either side of
AB, will then determine points through which a perpendicular
line to AB passes. An orthogonal grid is created by similarly constructing a series of perpendicular lines to one of
the orthogonal line to AB previously drawn.

7. The exact position of Teo 6 being unknown, Chiu and
Morrison (1980) proposed a location for this pecked cross.
Their explanation for the orientation and planning of Teotihuacan rests entirely on this hypothetical location of Teo 6,
and as they acknowledged, "two field bearings will test the
entire proposal" (Chiu and Morrison 1980, p.S63), the bearings toward Teo 1 and Mound Xalpan, once the accurate location of Teo 6 will be established.

8. In their search for an identical alignment to a celestial body from Teo 1 - Teo 5 and from Teo 6, Chiu and Morrison (1980, p.S60) eliminated the Pleiades as a possible candidate because of its poor distinction: "the first alignment
is not acceptable at a reasonable date and the second one is
less than striking with respect to any horizon feature".

9. The lines not following a regular 322 m interval were
in any case multiples of this unit of measure.

ACKNOWLEDGEMENT
    I wish to express to Antonio Mazza my sincere gratitude
for his valuable assistance.

References

AGI (Archivo General de Indias), Justicia, leg. 151, ff 68r-
        75r; leg. 157, ff 41v-45v.

Anderson, A.J.O.; F. Berdan; and J. Lockhart (eds). 1976.
        Beyond the Codices: The Nahua View of Colonial
        Mexico. Berkeley.

Anon. n.d.a. Libro de Guardias para la Corbeta Atrevida.
        Sig.755. Museo Naval, Madrid (Typescript).

Anon. n.d.b. Canto de Alegria. In: Virreinte de Mejico.
        Sig. 567. Museo Naval, Madrid (Typescript).

Applegate, R.B. n.d. Ineseño Chumash Grammar. Unpublished
        doctoral dissertation at the University of Cali-
        fornia, Berkeley.

Arochi, Luis. 1977. La Piramide de Kukulcan, su simbolismo
        solar. Editorial Orion, Mexico.

Ascher, Marcia and Robert Ascher. 1972. Numbers and rela-
        tions from ancient Andean quipus. Archive for
        History of Exact Sciences, 8: 288-320.

---. 1978. Code of the Quipu: Databook. University of
        Michigan Press, Ann Arbor.

---. 1980. Code of the Quipu: A Study in Media, Mathematics
        and Culture. University of Michigan Press, Ann
        Arbor.

Aubin, J.M.A. 1891. In: Documents pour servir à l'Histoire
        du Mexique: catalogue raisonné de la collection de
        M.E. Eugene Goupil, II, by E. Boban. Paris.

Aveleyra Arroyo, Luis; Manuel Maldonado; and Pablo Martinez
        del Rio. 1965. La Cueva de La Candelaria. V

Memoria del Instituto Nacional de Antropología e
Historia. Secretaria de Educación Pública,
México.

Aveni, Anthony F. 1975. Possible astronomical orientations
in ancient Mesoamerica. In: Archaeoastronomy in
Pre-Columbian America, pp. 163-190; ed. by A.F.
Aveni. University of Texas Press, Austin.

---. 1977. Concepts of positional astronomy employed in
ancient Mesoamerican architecture. In: Native
American Astronomy, pp. 3-19; ed. by A.F. Aveni.
University of Texas Press, Austin.

---. 1980. Skywatchers of Ancient Mexico. University of
Texas Press, Austin.

---. 1982. Archaeoastronomy in the Maya region: 1970-1980.
In: Archaeoastronomy in the New World, pp. 1-30;
ed. by A.F. Aveni. Cambridge University Press,
Cambridge.

---, and Robert M. Linsley. 1972. Mound J, Monte Alban:
possible astronomical orientation. American
Antiquity, 37: 528-531.

---; Sharon L. Gibbs; and Horst Hartung. 1975. The Caracol
tower at Chichen Itza: an ancient astronomical
observatory? Science, 188: 977-985.

---; Horst Hartung; and Beth Buckingham. 1978. The pecked
cross symbol in Mesoamerica. Science, 202: 267-
279.

---, and Horst Hartung. 1981. The observation of the sun at
the time of passage through the zenith in Mesoamerica. Journal for the History of Astronomy
(Archaeoastronomy 3), 12: S52-S70.

---, and Horst Hartung. 1982. Precision in the layout of
Maya architecture. Annals of the New York Academy

of Sciences, 385: 63-80.

Baillargeon, R.; G. Noelting; L-J. Dorais; and B. Saladin
d'Anglure. 1977. Aspects sémantiques et structu-
raux de la numération chez les Inuit. Etudes
Inuit, 1: 93-128.

Baraga, F. 1878. A Theoretical and Practical Grammar of the
Otchipwe Language. Montreal.

Barker, James. 1953. Memoria sobre la cultura de los
Guaika. Boletín Indigenista Venezolano, 1:
433-489. (HRAF translation).

Barnum, Francis. 1901. Grammatical Fundamentals of the
Inuit Language. Boston.

Bayly, William. 1776-1779. A Log and Journal Kept On Board
H.M. Sloop Discovery by William Bayly, August 1,
1776 to December 3, 1779. Alexander Turnbull
Library, Wellington, New Zealand.

Becher, Hans. 1960. Die Surara and Pakidai, zwei Yanoami-
Stämme in Nordwestbrasilien. Hamburg, Museum für
Völkerkinde, Mitteilungen, 26: 1-133. (HRAF
translation).

Beeler, Madison S. 1964. Ventureño numerals. In: Studies
in Californian Linguistics, pp. 13-18; ed. by W.
Bright. University of California Publications in
Linguistics, Vol. 34. Berkeley.

---. 1967. The Ventureño Confesionario of José Señán,
O.F.M. University of California Publications in
Linguistics, Vol. 47. Berkeley.

---. 1976. Barbareno Chumash grammar: a Farrago. In:
Hokan Studies (Papers from the First Conference on
Hokan Languages, held in San Diego, California,
April 23-25, 1970), pp. 251-269; ed. by M. Langdon
and S. Silver. The Hague -- Paris.

Berlin, Heinrich. 1968. The tablet of the 96 glyphs at
    Palenque, Chiapas, Mexico. Middle American
    Research Institute, Publication 26, pp. 135-149.
    Tulane University, New Orleans.

Bernal, Ignacio and Miguel Leon-Portilla. 1974. Historia de
    México, Vol. 2. Salvat Editores, S.A.,
    Barcelona.

Bills, Garland D.; Bernardo C. Vallejo; and Rudolph C.
    Troike. 1969. An Introduction to Spoken Bolivian
    Quechua. Institute of Latin American Studies.
    University of Texas Press, Austin.

Blom, Frans. 1935. The Pestac stela. Maya Research, 2:
    190-191.

Brabant, Augustin J. n.d. Indian Language Materials Rela-
    tive to the West Coast of Vancouver Island in the
    Archives of the Roman Catholic Diocese of Vic-
    toria. A.M.D.G. Dictionary of the Hesquiats or
    Nootka Language, 9 March 1911, British Columbia
    Archives.

Broda de Casas, Johanna. 1969. The Mexican calendar as com-
    pared to other Mesoamerican systems. Acta Ethno-
    logica et Linguistica, 15. Wien.

Burney, James. n.d. Log of the Discovery. Adm 51/4528.
    Public Record Office, London.

Cajori, Florian. 1919. The controversy on the origin of our
    numerals. The Scientific Monthly, November, pp.
    458-464.

Carlson, John B. 1976. The case for geomagnetic alignments
    of pre-Columbian Mesoamerican sites; the Maya.
    Paper presented at the 2nd Cambridge Symposium on
    Recent Research in Mesoamerican Archaeology, Cam-
    bridge, England.

---. 1977. The case for geomagnetic alignments of pre-
     Columbian Mesoamerican sites -- the Maya.
     Katunob, 10 (2): 67-88.

Carpenter, E. 1973. Eskimo Realities. New York.

Caso, Alfonso. 1958. The Aztecs: People of the Sun. Uni-
     versity of Oklahoma Press, Norman.

Castillo F., V.M. 1972. Unidades nahuas de medida. Estu-
     dios de Cultura Náhuatl, 10: 195-223. Mexico.

Chiu, B.C. and Philip Morrison. 1980. Astronomical origin
     of the offset grid at Teotihuacan. Journal for
     the History of Astronomy (Archaeoastronomy 2), 11:
     S55-S64.

Clancy, Flora. 1977. Relief carved monuments of Tikal: a
     possible geometry. Unpublished manuscript, Col-
     gate University.

Clarkson, Persis B. 1978. Classic Maya pictorial ceramics:
     a survey of content and theme. In: Papers on the
     Economy and Archictecture of the Ancient Maya, pp.
     86-141; ed. by R. Sidrys. Monograph 8, Institute
     of Archaeology, University of California, Los
     Angeles.

Cline, H.F. 1966. The Oztoticpac Lands Map of Texcoco,
     1540. Quarterly Journal of the Library of Con-
     gress, 23: 77-115.

---. 1968. The Oztoticpac Lands Map of Texcoco, 1540: Fur-
     ther Notes. 37th International Congress of Ameri-
     canists: Actas y Memorias, 3: 119-138. Buenos
     Aires.

Closs, Michael P. 1977a. The date reaching mechanism in the
     Venus Table of the Dresden Codex. In: Native
     American Astronomy, pp. 89-99; ed. by A.F. Aveni.
     University of Texas Press, Austin.

---. 1977b. The nature of the Maya chronological count. American Antiquity, 42: 18-27.

---. 1978. The Initial Series on Stela 5 at Pixoy. American Antiquity, 43: 690-694.

---. 1979. An important Maya inscription from the Xcalumkin area. mexicon, 1: 44-46.

---. 1982. On a Classic Maya accession phrase and a glyph for "rulership". mexicon, 4: 47-50.

---. 1983. A truncated Initial Series from Xcalumkin. American Antiquity, 48: 115-122.

Códice de Santa María Asunción, Apeo y Deslinde de Tierras (de los Terrenos) de Santa María Asunción. Ms. 1497bis. Biblioteca Nacional de México.

Codex Mariano Jimenez. 1967. Nómina de tributos de los pueblos Otlazpan y Tepexic, 1549; Códice Mariano Jimenez. Instituto Nacional de Antropología e Historia, Mexico.

Codex Vergara. Ms. Mex. 37-39. Bibliothèque Nationale de Paris.

Coe, Michael D. 1976. Early steps in the evolution of Maya writing. In: Origins of Religious Art and Iconography in Preclassic Mesoamerica, pp. 109-122; ed. by H.B. Nicholson. Los Angeles.

---, and Elizabeth P. Benson. 1966. Three Maya Relief Panels at Dumbarton Oaks. Studies in Pre-Columbian Art and Archaeology, No.2. Dumbarton Oaks, Washington, D.C.

Compte, F.M. de. 1885. Varones Ilustres de la Orden Seráfica en el Ecuador desde a fundación de Quito hasta nuestros dias. Quito.

Cook, James. 1967. The Journey of Captain James Cook on His

Voyage of Discovery. Vol. 3, Ots 1-2, The Voyage
of the Resolution and Discovery, 1776-1780; ed. by
J.C. Beaglehole. Hakluyt Society, Cambridge.

Copway, George (Kahgegagahbowh). 1851. The Traditional
History and Characteristic Sketches of the Ojibway
Nation. Boston.

Cortés, Hernán. 1538. Carta al consejo de Indias. In:
Colección de documentos inéditos, relativos al
descubrimiento..., sacados de los archivos del
reino, y muy especialmente del de Indias, 3:
535-545. Madrid.

Cortés, Martín. 1563. Carta al Rey D. Felipe II, 10 de
Octubre. In: Colección de documentos inéditos,
relativos al descubrimiento...., sacados de los
archivos del reino, y muy especialmente del de
Indias, 4: 440-462. Madrid.

Curtis, Edward S. 1916. The North American Indian, Vol. 11.
Ed. F.W. Hodge. [Johnson Reprint Corporation, New
York, 1970.]

---. 1926. The North American Indian, Vol. 15. [Johnson
Reprint Corporation, New York, 1970.]

Cushing, Frank H. 1982. Manual concepts: a study of the
influence of hand usage on culture-growth. American Anthropologist, 5: 289-317.

Datta, B. and A.N. Singh. 1962. History of Hindu Mathematics (2 volumes in 1). Bombay.

Davies, Nigel. 1977. The Toltecs, until the fall of Tula.
University of Oklahoma Press, Norman.

Day, C.L. 1967. Quipus and Witches Knots. University of
Kansas Press, Lawrence.

Denny, J. Peter. 1981. The logical semantics of 'only':
tuaq, innaq, and tuinnaq. Inuit Studies, 5

(Supplementary Issue): 115-124.

Densmore, Frances. 1929. Chippewa customs. Smithsonian Institution, Bureau of American Ethnology, Bulletin 86. Washington, D.C.

Dewdney, Selwyn. 1975. The Sacred Scrolls of the Southern Ojibway. University of Toronto Press, Toronto.

Dibble, Charles E. 1963. Historia de la Nación Mexicana, Reproducción a todo color del 1576 (Códice Aubin), Versión Paleográfica y Traducción Directa del Náhuatl. Colección Chimalistac, Edicione Jose Porrua Turanzas.

Dixon, Roland B. and A.L. Kroeber. 1907. Numeral systems of the languages of California. American Anthropologist, 9: 663-690.

Dorsey, James O. and John R. Swanton. 1912. A dictionary of the Biloxi and Ofo languages. Smithsonian Institution, Bureau of American Ethnology, Bulletin 47. Washington, D.C.

Drewitt, Bruce. n.d. Measurement units and building axes at Teotihuacan. Unpublished manuscript, University of Toronto.

Drucker, Philip. 1951. The Northern and Central Nootkan Tribes. Smithsonian Institution, Bureau of American Ethnology, Bulletin 144. Washington, D.C.

---. n.d. Nootka Field Notebooks, August 1935 - December 1936. 15 Vols. Bureau of American Ethnology, Manuscript Collection, No. 4516, Pt. 23. Smithsonian Institution. National Anthropological Archives.

Eber, Dorothy. 1972. Eskimo art: looking for the artists of Dorset. The Canadian Forum, 52: 12-16.

Edmonson, Munro S. 1971. The Book of Counsel: The Popol Vuh

of the Quiche Maya of Guatemala. Middle American
Research Institute, Pub. 35, Tulane University.
New Orleans.

Eells, W.C. 1913. Number systems of the North American
Indians. The American Mathematical Monthly, 20:
263-299.

Fettweis, Martine. 1973. Propections archaeologiques sur la
cote du Quintana Roo, Mexique 1970. Contributions
à l'étude de l'architecture et de la peinture
murale. Thèse présentée à l'Université de Lou-
vain, Belgique.

Fettweis-Vienot, Martine. 1980. Las pinturas murales de
Coba. Boletín de la Escuela de Ciencias Antropo-
lógicas de la Universidad de Yucatán, 7(40): 2-50.

Folan, William J. 1970. Yuquot, Where the Wind Blows from
All Directions: The Ethnohistory of the Nootka
Sound Region. Revised, 1976. Unpublished manu-
script on file in Parks Canada, Department of
Indian and Northern Affairs, Ottawa.

---, and Antonio Ruiz Perez. 1980. The diffusion of astronomical knowledge in greater Mesoamerica.
Archaeoastronomy, III (3): 20-25.

Fox, James A. and John S. Justeson. 1984. Polyvalence in
Mayan hieroglyphic writing. In: Phoneticism in
Mayan Hieroglyphic Writing, pp. 17-76; ed. by John
S. Justeson and Lyle Campbell. Institute for
Mesoamerican Studies, State University of New York
at Albany, Publication 9. Albany.

Fulton, Charles C. 1948. Did the Maya Have a Zero? Carnegie
Notes on Middle American Archaeology and Ethnology, 3: 233-239. Washington, D.C.

Gallatin, A. 1845. Notes on the semi-civilized nations of

Mexico, Yucatan, and Central America. Transactions of the American Ethnological Society, 1:
1-352.

Gatschat, Albert S. and John R. Swanton. 1932. A dictionary
of the Atakapa language. Smithsonian Institution,
Bureau of American Ethnology, Bulletin 108.
Washington, D.C.

Gay, J. and M. Cole. 1967. The New Mathematics and an Old
Culture. Holt, Rinehart and Winston, New York.

Ghinassi, J. 1938. Gramatica Teorico-Practica y Vocabulario
de la Lengua Jibara. Talleres Graficos de Educa-
cion, Quito.

Gibson, C. 1964. The Aztecs Under Spanish Rule. Stanford
University Press, Stanford.

Graham, Ian and Eric Von Euw. 1977-1979. Corpus of Maya
Hieroglyphic Inscriptions, Vol. 3, Parts 1-2.
Peabody Museum of Archaeology and Ethnology, Har-
vard University, Cambridge.

Graham, John A. 1972. The Hieroglyphic Inscriptions and
Monumental Art of Altar de Sacrificios. Papers of
the Peabody Museum of Archaeology and Ethnology,
Vol. 64, No. 2. Harvard University, Cambridge.

---. 1977. Discoveries at Abaj Takalik, Guatemala. Archaeology, 30: 196-197.

Guerra, Francisco. 1969. Aztec science and technology.
History of Science, 8: 32-52.

Hallowell, A.I. 1942. Some psychological aspects of measurement among the Saulteaux. Reprinted in Culture
and Experience; 1955. New York.

Hallpike, C.R. 1979. The Foundations of Primitive Thought.
Clarendon Press, Oxford.

Harner, M.J. 1972. The Jivaro, People of the Sacred

Waterfalls. Doubleday Natural History Press, New York.

Hartung, Horst. 1977. Ancient Maya architecture and planning: possibility and limitations for astronomical studies. In: _Native American Astronomy_, pp. 111-129; ed. by A.F. Aveni. University of Texas Press, Austin.

Harvey, Herbert R. and Barbara J. Williams. 1980. Aztec arithmetic: positional notation and area calculation. _Science_, 210: 499-505.

---. 1981. L'arithmetique azteque. _La Recherche_, 126: 1068-1081.

Hassel, J.M. von. 1902. Vocabulario Aguaruna. In: _Boletín de la Sociedad Geografica de Lima_, Anno 12, XII: 73-86.

Haswell, Robert. 1941. Voyages of the _Columbia_ to the Northwest Coast, 1787-1790 and 1790-1793. In: _Collections of the Massachusetts Historical Society, Vol. 79_; ed. by F.W. Howay. Boston.

Hébert-Stevens, Francois. 1972. _L'Art Ancien de l'Amérique du Sud_. Arthaud, Paris.

Heizer, Robert F. 1955. California Indian Linguistic Records, The Mission Indian Vocabularies of H.W. Henshaw. _Anthropological Records_, 15(2).

---, and T.R. Hester. 1978. Two petroglyph sites in Lincoln County, Nevada. In: _Four Rock Art Studies_, pp. 1-44; ed. by W. Clewlow. Ballena Press, Socorro, New Mexico.

Heyden, Doris. 1975. An interpretation of the cave underneath the Pyramid of the Sun in Teotihuacan, Mexico. _American Antiquity_, 40: 131-147.

Hickerson, Harold. 1963. Notes on the post-contact origin

of the Midewiwin. Ethnohistory, 9: 404-423.

---. 1970. The Chippewa and their Neighbors: A study in Ethnohistory. New York.

Hoffman, W.J. 1891. The Midé'wiwin or "Grand Medicine Society" of the Ojibwa. Smithsonian Institution, Bureau of American Ethnology, 7th. Annual Report. Washington, D.C.

Holmberg, Allan R. 1950. Nomads of the long bow; the Siriona of Eastern Bolivia. Smithsonian Institution, Institute of Social Anthropology, Pub. No.10. Washington, D.C.

Ixtlilxóchitl, Fernando de Alva. 1977. Obras Históricas 2. Mexico.

Jewitt, John Rogers. 1807. A Journal Kept at Nootka Sound. Printed for the author, Boston.

Johnston, Basil. 1976. Ojibway Heritage. Toronto.

Jones, William and Truman Michelson. 1903. Algonquian (Fox). In: Handbook of American Indian Languages, Part 1. Smithsonian Institution, Bureau of American Ethnology, Bulletin 40. Washington, D.C.

Karsten, R. 1935. The Head-Hunters of Western Amazonas. The Life and Culture of the Jibaro Indians of Eastern Ecuador and Peru. Societas Scientiarum Fennica, Helsingfors.

Kay, P. 1979. The Role of Cognitive Schemata in Word Meaning: Hedges Revisited. MS, Language Behavior Research Laboratory, University of California, Berkeley.

Kelley, David H. 1976. Deciphering the Maya Script. University of Texas Press, Austin.

Kirkland, F. and W.W. Newcomb. 1967. The Rock Art of Texas

Indians. University of Texas Press, Austin.

Klar, Kathryn A. 1980. Northern Chumash numerals. In: American Indian and Indoeuropean Studies (Festschrift Beeler), pp. 113-119. The Hague.

Knipe C. 1868. Some Account of the Tahkaht Language as Spoken by Several Tribes on the Western Coast of Vancouver Island. n.p. London.

Kroeber, A.L. 1923. Anthropology [with a supplement, 1933; revised version, 1948]. New York.

---. 1925. Handbook of the Indians of California. Smithsonian Institution, Bureau of American Ethnology, Bulletin 78. Washington, D.C.

---, and George W. Grace. 1960. The Sparkman Grammar of Luiseño. University of California Publications in Linguistics, 16. Berkeley.

---. 1963. Yokuts Dialect Survey. Anthropological Records, 11(3).

La Flesche, Francis. 1932. A dictionary of the Osage language. Smithsonian Institution, Bureau of American Ethnology, Bulletin 109. Washington, D.C.

Lameiras, Brigitte B. 1974. Terminología Agrohidráulica Prehispánica Nahua. Instituto Nacional de Antropología e Historía, Colección Científica 13. Mexico.

Landa, Diego de. 1941. Landa's relación de las cosas de Yucatan. A translation with notes by Alfred M. Tozzer. Papers of the Peabody Museum of American Archaeology and Ethnology, Harvard University, 18. Cambridge.

Landes, Ruth. 1968. Ojibwa Religion and the Midéwiwin. The University of Wisconsin Press, Madison.

Leander, B. 1967. Códice de Otlazpan, Instituto Nacional de

Antropología e Historia, Serie Investigaciones 13.
Mexico.

León-Portilla, Miguel. 1974. Historia de México, Vol. 3.
Salvat Editores, S.A., Barcelona.

Lévi-Strauss, Claude. 1968. The Savage Mind [translated
from the French]. University of Chicago Press.

Levy-Bruhl, L. 1912. Les Fonctions Mentales dans les Soci-
étés Inférieures. Paris.

Lizardi Ramos, Cesar. 1962. El 'cero' Maya y su función.
Estudios de Cultura Maya, 2: 343-353.

Locke, Leland L. 1932. The ancient Peruvian abacus.
Scripta Mathematica, 1: 37-43.

---. 1923. The Ancient Quipu or Peruvian Knot Record. The
Museum of Natural History, New York.

Loeb, Edwin M. 1926. Pomo folkways. University of California Publications in American Archaeology and
Ethnology, 19. Berkeley.

---. 1964. Die Institution des Sakralen Königtums.
Paideuma, 10: 102-114.

Lounsbury, Floyd G. 1973. On the derivation and reading of
the 'Ben-Ich' prefix. In: Mesoamerican Writing
Systems, pp. 99-143; ed. by Elizabeth P. Benson.
Dumbarton Oaks, Washington, D.C.

---. 1974. The inscription of the sarcophagus lid at
Palenque. In: Primera Mesa Redonda de Palenque,
Part 2, pp. 5-19; ed. by Merle Greene Robertson.
The Robert Louis Stevenson School, Pebble Beach,
California.

---. 1976. A rationale for the initial date of the Temple
of the Cross at Palenque. In: The Art, Iconography and Dynastic History of Palenque, Part 3, pp.
211-224; ed. by Merle Greene Robertson. The

Robert Louis Stevenson School, Pebble Beach, California.

---. 1978. Maya numeration, computation, and calendrical astronomy. Dictionary of Scientific Biography, Vol.15, Supplement 1, pp. 759- 818. New York.

Manya A., Juan Antonio. 1972. Hablando Quechua con el Pueblo. Instituto Linguistico y Folklorico Quechua y Aimara, Cusco, Peru.

Marcus, Joyce. 1983. The conquest slabs of Building J, Monte Alban. In: The Cloud People: Divergent Evolution of the Zapotec and Mixtec Civilizations; ed. by K.F. Flannery and J. Marcus. New York.

Marshack, Alexander. 1972. Roots of Civilization: Cognitive Beginnings of Man's First Art Symbol and Notation. New York.

Mathews, Peter and Linda Schele. 1973. Lords of Palenque -- The Glyphic Evidence. In: Primera Mesa Redonda de Palenque, Part 1, pp. 63-75; ed. by Merle Greene Robertson. The Robert Louis Stevenson School, Pebble Beach.

Matteson, Esther. 1965. The Piro (Arawakan) language. University of California Publications in Linguistics, 42. Berkeley.

Maudslay, Alfred P. 1889-1902. Archaeology. Biologia Centrali-Americana (5 vols). London.

McGee, W.J. 1900. Primitive numbers. Smithsonian Institution, Bureau of American Ethnology, 19th Annual Report, pp. 821-851. Washington, D.C.

Menninger, Karl. 1969. Number words and number symbols. English translation by Paul Broneer. Cambridge.

Miller, Arthur C. 1973. The Mural Painting of Teotihuacan. Dumbarton Oaks, Washington, D.C.

Moffat, Hamilton. n.d. Letterbook of Hamilton Moffat. Fort
    Rupert, Fort Simpson and Fort Kamloops 1857-1867.
    British Columbia Archives, A/B/20/R2A.
Morley, Sylvanus G. 1915. An Introduction to the Study of
    the Maya Hieroglyphs. Smithsonian Institution,
    Bureau of American Ethnology. Bulletin 57.
    Washington, D.C.
Mountford, Cheles P. 1964. Pinturas Australianas Abori-
    genes. Editorial Hermes, México.
Moziño Suarez de Figueroa, José Marino. 1913. Noticias de
    Nutka, Diccionario de la Lengua de los Nutkenses y
    Descripcion del volcan de Tuxtla: por Joseph
    Marino Moziño Suarez de Figueroa. Precididos de
    una noticia acerca del br. Moziño y de la expedi-
    cion científica del siglo XVIII por Alberto M.
    Carreno. La Secretaria de Fomento, Madrid.
    Translated by Secretary of State and revised by
    William J. Folan.
Murray, W. Breen. 1979a. Interpretive perspectives on the
    rock art of Northeast and North Central Mexico,
    with special reference to Neuvo Leon. South Texas
    Journal of Research and the Humanities, 3(1):
    27-52.
---. 1979b. Description and analysis of a petroglyphic
    tally count stone at Presa de La Mula, N.L.,
    Mexico. mexicon, 1: 7-9.
---. 1982. Rock art and site environment at Boca de Potre-
    rillos, N.L., Mexico. American Indian Rock Art,
    8.
---. 1984. Numerical characteristics of three engraved
    bison scapulae from the Texas Gulf Coast.
    Archaeoastronomy, 7: 82-88.

———. 1985. Petroglyphic counts at Icamole, Neuvo Leon
        (Mexico). Current Anthropology, 26: 276-279.

Nance. C. Roger. 1971. The Archaeology of La Calzada: a
        Stratified Rock Shelter Site, Sierra Madre Orien-
        tal, Nuevo Leon, Mexico. Unpublished Ph.D. Dis-
        sertation. Department of Anthropology, University
        of Texas, Austin.

Nekapmarar' nakurustai. 1976-1978. [Four booklets]. Fede-
        racion de Centros Shuar, Escuelas radiofonicas,
        Sucua.

Nelson, Edward W. 1899. The eskimo about Bering Strait.
        Smithsonian Institution, Bureau of American Ethno-
        logy, 18th Annual Report, Part 1. Washington,
        D.C.

Neugebauer, Otto. 1957. The Exact Sciences in Antiquity.
        Brown University Press, Providence.

Newman, S. 1944. Yokuts Language of California. Viking
        Fund Publications in Anthropology (2). New York.

Nicholson, Henry B. 1971. Religion in pre-hispanic Central
        Mexico. In: Handbook of Middle American Indians,
        10, pp. 395-446; ed. by G.F. Ekholm and I Bernal.
        University of Texas Press, Austin.

Novo y Colson, Pedro de. 1885. Politico-Scientific Voyage
        Round the World by the Corvettes Descubierta and
        Attrevida under the Command of the Naval Captains
        Don Alejandro Malaspina and Don Jose de Bustamente
        Guerra. 1784-1794. Published with an introduc-
        tion by Don Pedro de Novo y Colson, Lieutenant,
        Spanish Royal Navy, Correspondent of the Royal
        Academy of History. 3 Vols. Madrid (Typescript).
        British Columbia Archives. A/A/30/M29.

Orozco y Berra, Manuel. 1960. Historica Antigua y de la

Conquista de México. Ed. prep. por. A.M. Garibay
K. y M. Leon-Portilla, 4 vols. Editorial Porrua,
S.A., Mexico.

Papeles de la Embajada Americana. 3d Series, Exp. 2-7, Reg.
10-3, Doc 3. Museo Nacional de Antropología,
Archivo Histórico. Mexico.

Paso y Troncoso, Franciso del. 1912. Códice Kingsborough.
Memorial de los Indios de Tepetlaoztoc al monarca
español contra los encomenderos del pueblo.
Madrid.

Pellizzaro, S.M. 1969. Apuntes de Gramatica Shuar. Misio-
nes Salesianas, Quito.

Proskouriakoff, Tatiana. 1960. Historical implications of a
pattern of dates at Piedras Negras, Guatemala.
American Antiquity, 25: 454-475.

---. 1963. Historical data in the inscriptions of Yaxchilan. Part 1. Estudios de Cultura Maya, 3:
149-167.

---. 1964. Historical data in the inscriptions of Yaxchilan. Part 2. Estudios de Cultura Maya, 4:
177-201.

Raimondi, A. 1863. On the Indian tribes of Loreto. In:
Anthropology Review, 1: 35-43.

Reiss, W. 1880. Ein Besuch bei den Jivaros-Indianern. In:
Verhandlungen Gesellsschaft fur Erdkunde, VII:
325-337.

Rivet, P. 1907-1908. Les Indiens Jivaros. Etude Géographi-
que, Historique et Ethnographique. In: L'Anthro-
pologie, XVIII, (3-6): 333-368 and 583-618; XIX,
(1-3): 69-87 and 235-259.

Robertson, Merle Greene. 1977. Painting practices and their
change through time of the Palenque stucco

sculptors. In: Social Process in Maya Prehistory, pp. 297-326; ed. by N. Hammond. London.

———. 1983. The Sculpture of Palenque. Princeton University
Press, Princeton.

Rodriguez, A. 1969. A History of Mexican Mural Painting.
New York.

Ross, Kurt. 1978. Codex Mendoza, Aztec Manuscript. Produc-
tions Liber, S.A., CH-Fribourg. Spain.

Sanders, Frank J. 1977. The "twin-stelae" of Seibal.
American Antiquity, 42: 78-86.

Sapir, Edward. 1903. The Takelma language of southwestern
Oregon. In: Handbook of American Indian Languages, Part 2. Smithsonian Institution, Bureau
of American Ethnology, Bulletin 40. Washington,
D.C.

———, and M. Swadesh. 1960. Yana dictionary. University of
California Publications in Linguistics, 22.
Berkeley.

Scarborough, Vernon; Beverly Mitchum; Sorraya Carr; and David
Freidel. 1982. Two Late Preclassic ballcourts at
the lowland Maya center of Cerros, Northern
Belize. Journal of Field Archaeology, 9: 21-34.

Schele, Linda. 1977. Palenque: the house of the dying sun.
In: Native American Astronomy, pp. 42-56; ed. by
A.F. Aveni. University of Texas Press, Austin.

Schoolcraft, Henry R. 1851. History, Condition and Pros-
pects of the Indian Tribes of the United States.
6 volumes. Philadelphia.

Seidenberg, A. 1960. The diffusion of counting practices.
University of California Publications in Mathematics, 3: 215-300. Berkeley.

———. 1962. The ritual origin of counting. Archive for

History of Exact Sciences, 2: 1-40.

---. 1965. The sixty system of Sumer. Archive for History of Exact Sciences, 2: 436-440.

Seler, Eduard. 1904. The Mexican picture writings of Alexander von Humboldt. Smithsonian Institution, Bureau of American Ethnology, Bulletin 28. Washington, D.C.

Siméon, Rémi. 1977. Diccionario de la lengua nahuatl o mexicana. Josefina Oliva de Coll (trans). Mexico.

Smith, David E. 1923-1925. History of Mathematics, 2 vols. Boston.

Spencer, Charles S. 1982. The Cuicatlan Canada and Monte Alban: A Study of Primary State Formation. Academic Press, Orlando.

Sproat, Gilbert Malcolm. 1868. Scenes and Studies of Savage Life by Gilbert Malcolm Sproat. Smith, Elder and Co., London.

Steinen, Karl von den. 1894. Unter den naturvölkern zentral-Brasiliens. Reiseschilderung und ergelnisse der zweiten Schingú-expedition 1887-1888. Berlin. (HRAF translation).

Strange, James. 1928. James Strange's Journal and Narrative of the Commercial Expedition from Bombay to the North-West Coast of America, together with a chart showing the tract of the Expedition. Government Press, Madras.

Sullivan, Thelma D. 1963. Náhuatl proverbs, conundrums and metaphors collected by Sahagún. Estudios de Cultura Náhuatl, 4: 93-178.

---. 1976. Compendio de la Gramática Náhuatl. Universidad Nacional Autónoma de México, Instituto de

Investigaciones Históricas, México.

Taladoire, Eric. 1979. Orientation of ball-courts in Meso-america. Archaeoastronomy, 2: 12-13.

Taylor, J. Garth. 1980. Canoe Construction in a Cree Cultural Tradition. National Museum of Man, Ottawa.

Teeple, John E. 1931. Maya astronomy. Contributions to American Anthropology and History, No. 2. Carnegie Institution of Washington, Publication 403. Washington.

Teeter, Karl V. 1964. The Wiyot language. University of California Publications in Linguistics, 37. Berkeley.

Thomas, Cyrus. 1900. Numeral systems of Mexico and Central America. Smithsonian Institution, Bureau of American Ethnology, 19th. Annual Report, Part 2, pp. 853-955. Washington, D.C.

Thompson, J. Eric S. 1971. Maya Hieroglyphic Writing: An Introduction. University of Oklahoma Press, Norman.

---. 1972. A commentary on the Dresden Codex. Memoirs of the American Philosophical Society, Vol.93. Philadelphia.

Tomkins, Peter. 1976. Mysteries of the Mexican Pyramids. Harper and Row, New York.

Trumbull, J. Hammond. 1874. On numerals in American Indian languages, and the Indian mode of counting. American Philogical Association, Transactions and Proceedings, 5: 41-76.

Tylor, E.B. 1903 Primitive Culture: Researches into the Development of Mythology, Philosophy, Religion, Art and Custom. 2 Vols. (4th edition). John Murray, London. [1st Edition, 1871].

Van der Waerden, B.L. 1961. Science Awakening [2nd Edi-
        tion]. Groningen.

Vaillant, George C. 1950. The Aztecs of Mexico. Pelican
        Books.

Viana. n.d. Diario del viage explorador de las Corbetas
        Descubierta y Atrevida. Sig. 892. Museo Naval,
        Madrid (Typescript).

Villacorta, J. Antonio and Carlos A. Villacorta. 1930.
        Códices mayas, reproducidos y desarrollados.
        Guatemala.

Warren, William W. 1957. History of the Ojibway Nation.
        Collections of the Minnesota Historical Society,
        Vol. 5, St. Paul, 1885. Re-published Minneapolis.

Wassén, H. 1931. The ancient Peruvian abacus. Comparative
        Ethnological Studies, 9: 191-205.

Williams, Barbara J. 1980a. Náhuatl Soil Glyphs from the
        Códice de Santa María Asunción, Actes du XLII [e]
        Congrès International des Américanistes, 9:
        165-175. Paris.

---. 1980b. Pictorial Representation of Soils in the Valley
        of Mexico: Evidence from the Codex Vergara. Geo-
        science and Man, 21: 51-62.

---. 1984. Mexican Pictorial Cadastral Registers: An anal-
        ysis of the Códice de Santa María Asunción and the
        Codex Vergara. In: Explorations in Ethnohistory:
        Indians of Central Mexico in the Sixteenth Cen-
        tury, pp. 103-125; ed. by H.R. Harvey and H.J.
        Prem. Albuquerque.

Zimmerly, David W. 1979. Hooper Bay Kayak Construction.
        National Museum of Man, Ottawa.

Zorita, Alonzo de. 1963. Life and Labor in Ancient Mexico:
        The Brief and Summary Relation of the Lords of New

Spain. Benjamin Keen (trans). New Brunswick.

Zuidema, R. Thomas. 1964. The Ceque System of Cuzco: The
Social Organization of the Capital of the Inca.
International Archives of Ethnography, E.J. Brill,
Leiden.

---. 1977. The Inca calendar. In: Native American Astronomy, pp. 219-259; ed. by A.F. Aveni. University
of Texas Press, Austin.

---. 1982. The Sidereal Lunar calendar of the Incas. In:
Archaeoastronomy in the New World, pp. 59-106; ed.
by A.F. Aveni. Cambridge University Press,
Cambridge.

Contributors

Marcia Ascher
Department of Mathematics
Ithaca College
Ithaca, New York

Madison S. Beeler
Emeritus Professor
Department of Linguistics
University of California
Berkeley, California

Michael P. Closs
Department of Mathematics
University of Ottawa
Ottawa, Ontario

Maurizio Covaz Gnerre
Departmento de Lingüística
UNICAMP
São Paulo, Brazil

J. Peter Denny
Department of Psychology
University of Western Ontario
London, Ontario

William J. Folan
Director
Centro de Investigaciones
    Históricas y Sociales
Universidad del Sudeste
Campeche, Mexico

Herbert R. Harvey
Department of Anthropology
University of Wisconsin
Madison, Wisconsin

William Breen Murray
Facultad de Humanidades
Universidad de Monterrey
Monterrey, Mexico

Stanley E. Payne
Department of Mathematics
University of Colorado at Denver
Denver, Colorado

A. Seidenberg
Department of Mathematics
University of California
Berkeley, California

Francine Vinette
Department of Applied Mathematics
University of Waterloo
Waterloo, Ontario

Barbara J. Williams
Department of Geography
University of Wisconsin
Janesville, Wisconsin

E 59 .M34 N37 1986
109776
Native American mathematics

DATE DUE	
SEP 26 1995	

SUSAN COLGATE CLEVELAND
LIBRARY/LEARNING CENTER
COLBY-SAWYER COLLEGE
New London, New Hampshire 03257

GAYLORD   PRINTED IN U.S.A.